Building Failures

Building Failures

A guide to diagnosis, remedy and prevention

Third edition

Lyall Addleson

Butterworth Architecture
An imprint of Butterworth-Heinemann Ltd
Linacre House, Jordan Hill, Oxford OX2 8DP

PART OF REED INTERNATIONAL BOOKS

OXFORD   LONDON   BOSTON
MUNICH   NEW DELHI   SINGAPORE   SYDNEY
TOKYO   TORONTO   WELLINGTON

First published by Architectural Press Ltd 1982
Reprinted 1983
Second edition published by Butterworth Architecture 1989
Third edition 1992

© Butterworth-Heinemann Ltd 1989, 1992

1064498 9

British Library Cataloguing in Publication Data
Addleson, Lyall
   Building failures.
   1. Great Britain. Building components. Failue
   I. Title
   691

ISBN 0 7506 0226 0

Library of Congress Cataloguing in Publication Data
Addleson, Lyall
   Building failures: a guide to diagnosis, remedy and
   prevention/Lyall Addleson
   p. 297 cm.
   Includes bibliographies and index.
   ISBN 0 7506 0226 0
   1. Building failures. 2. Buildings – Repair and
   reconstruction.
   I. Title.
   TH441.A33   88-35238
   690'21-dc19

Composition by Genesis Typesetting, Laser Quay, Rochester, Kent
Printed in Great Britain

21/93

Foreword to the third edition

In this third edition errors in the second edition have been corrected and all sections have been further updated. The selective list of organisations that undertake research or testing that first appeared in the second edition have been revised and expanded. In particular a more considered review is given of BS 5250: 1989 BS Code of practice for *Control of condensation in buildings* and the complementary BRE Report, *Thermal insulation: avoiding the risks.*

Some failures that were occurring in small numbers in the 1980s are occurring more commonly, or nearly so. These include defects related to single roofing membranes, cavity fill, profiled metal roof cladding and some types of insulated and other wall claddings. My selection and treatment of these is based largely on my own experience and the emphasis I might give reflects this.

The gathering together of a wide range of authoritative references remains a characteristic and important feature of the book. I have retained the presentation of relevant advice from outside sources in succinct and distilled form, the references being closely related to those parts to which they refer. As in the first edition, I have directed readers to specific parts and/or page numbers of the references.

Inevitably some of the references and information given about recommended practice, precautions, rules and products will be out of date or superseded by the time the book is published. The references and product information were correct as at the end of December 1991. Some references already out of date or a little old in the tooth have nevertheless been retained if they contain fundamental concepts or principles that should still be of value. References to my *Materials for building*, Vols 1–3 have mostly been superseded by *Performance of materials in buildings* (Lyall Addleson and Colin Rice, Butterworth-Heinemann, 1992). This new book also has a full treatment of *Principles for building*, first introduced in the second edition in Study 1, *Lessons.*

There are a number of BS Codes of Practice, such as those related to built-up felt and asphalt, whose revision is long overdue. In the meantime, the PSA's *Technical guide to flat roofing* should fill the gap. New codes are required on, for example, profiled metal roof cladding and wall cladding in general. In the meantime the first authoritative guidance on rainscreen cladding by Anderson and Gill for CIRIA, has appeared. Readers ought to check the most up-to-date BSs in the annual BSI *Standards Catalogue* including in particular changes from British to European standards. Similarly, they should also check the currency of other publications given, perhaps going first to BBA's monthly *Index* and Building File's quarterly *Index* or annual publications such as BRE's *Information Directory, Specification* or the *AJ's Directory of Information.*

It is most important that readers recognise that this book is a *guide* and no more. It is certainly not a copybook. The diagrams in Part 2 illustrate schematically the relevant constructional details. Those details are general and the principles illustrated or described in the text must therefore be interpreted in relation to the specific construction and conditions found on site in any particular case.

Importantly, the principles discussed and the technical or other advice given in this book are the product of careful study of the problems involved and the present state of the art. The publishers and I cannot guarantee the suitability of the solutions described for individual problems, and we shall be under no legal obligation of any kind in any respect of or arising out of the form or contents of this book or any error therein, or reliance of any person thereon.

Lyall Addleson
London, January 1992

Acknowledgements

The author repeats his gratitude to all at the Building Research Establishment and formerly at the disbanded Greater London Council for their patience, guidance and help during the preparation of the original articles, and subsequently and for supplying photographs and other illustrative material. The sources of all theses are acknowledged where they occur. The sources of facsimile and other reproductions are similarly acknowledged. The author and publisher are grateful for permission to use parts of the relevant publications.

In particular, extracts from British Standards are reproduced with the permission of the British Standards Institution: see page 23, 3.02 for address for copies. Building Research Establishment Crown Copyright material is reproduced by permission of the Controller of HMSO.

A few people deserve special mention and the author's gratitude: to Maritz Vandenberg, technical editor of the *AJ* when the original articles were published, for his optimism, enthusiasm and guidance; to Cluny Gillies, then at Butterworth Architecture, for his editorship of both earlier editions; to Colin Rice, co-author of *Performance of materials in buildings*, for his valuable assistance in updating, revising and checking of references for the first edition; and to Geoffrey Browne, the author's former clerical assistant for his thorough proof-reading and assembly of all the revised material for the second edition.

The author has special words of thanks: to all the readers of the *AJ* who drew attention to errors or omissions in the articles; to reviewers of the first edition for their helpful comments; to Tom Lawson for his expert advice on wind-related problems; to other experts for the 'opposition' for keeping me on my toes; and to my clients and their lawyers (solicitors, barristers and Counsel) for their patience while he tries to answer some of their more difficult questions. And last but not least to the author's wife for her encouragement and forbearance over many years.

Contents

Part 1: The approach to building failures

Study 1
Lessons

1 Lessons from failures

1.01 Pressure of economics

'The weaknesses of well tried materials are known and allowed for in good design; but even the best designer may have to use materials with known shortcomings because, for a particular purpose, there exists at the time no economically acceptable alternative.' (NBS Special Report 33, *A qualitative study of some buildings in the London area*, London, HMSO, 1964, p.1.)

With hindsight it is now patently obvious that, from 1945 to 1970, far too many new materials and building techniques, most of whose shortcomings were not necessarily understood, were used mostly because no economically acceptable alternatives seemed to exist. The building industry was then under intense pressure (the social, economic and political forces alone have not been inconsiderable) to produce without adequate resources an unprecedentedly large number of buildings with quite different performance requirements from those previously encountered.

1.02 Risks of innovation

A characteristic and alarming feature of post-war building failures is their *scale* (number, variety and cost) and *frequency*. Building failures are, of course, not unknown in the history of building and it it unlikely that they can be avoided completely. The next generation of buildings may have fewer failures if, as Gordon Wigglesworth* has suggested, there is less indulgence of untested innovation, and if, as Ian Freeman† asserts, far more reliance is placed on available recommendations for good practice. Although studies in this book go some way in directing architects to the latter, the success of building from a technological point of view also depends on the extent to which architects apply the following lessons.

1.03 A new look at principles, rules and precautions

Background

Many building failures could be avoided, or at least the worst of their effects reduced considerably, if principles and rules applicable to building could be differentiated. So far, the two are invariably combined and loosely referred to as 'Principles of building'. To these, precautions that should be taken in design and assembly are added.

Rules and precautions are the product of necessity and practicability. They emerge from experience in use in assembling or in the performance of buildings and relate to what is sometimes referred to as 'buildability'. The rules and precautions may (and do) change with experience; the principles *applicable* to buildings (described as 'Principles for building') do not.

It should be easier to identify aspects of detailed

* See *AJ* 11.8.76, pp. 252–253.
† See *AJ* 5.2.75, pp. 303–308, CI/SfB (9–) (S).

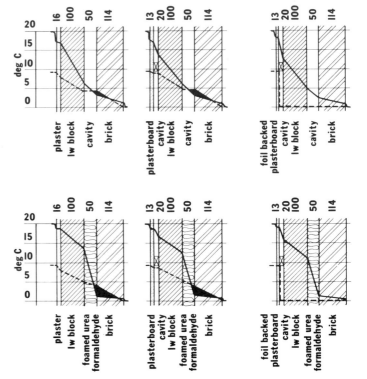

1 *Hazards of multi-layer construction: the standard cavity wall with alternative types and positions of insulation applied to it, giving different risks of interstitial condensation. Using the now outdated temperature gradient technique, condensation is shown (vividly) to occur in the solid black 'zone' (i.e. where the structural temperature gradient (solid line) falls below the dewpoint (dotted line)). It is now believed that condensation occurs at particular interfaces ('planes'), the location of which can be predicted: see Study 20 Condensation.*

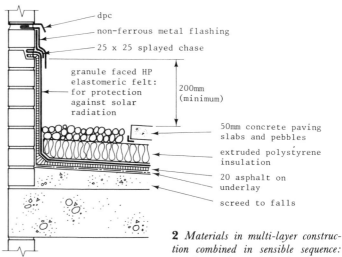

dpc
non-ferrous metal flashing
25 x 25 splayed chase
granule faced HP elastomeric felt: for protection against solar radiation
200mm (minimum)
50mm concrete paving slabs and pebbles
extruded polystyrene insulation
20 asphalt on underlay
screed to falls

2 *Materials in multi-layer construction combined in sensible sequence: weatherproof extruded polystyrene insulation placed above membrane, rather than below, thus reducing failure risk.*

constructional design that may lead to defects or failures if there were only a few principles to remember. Then it would be necessary simply to check the extent to which the relevant rules and precautions have changed in time.

In this study only a brief explanation is given. The full treatment of this approach is in *Performance of materials in buildings* by Lyall Addleson and Colin Rice, Butterworth-Heinemann, 1992, *1 Principles for building*, pages 15–34.

Principles for building
Summarized, these are:

Constraints

1 HIGH → LOW	• Gravity (water flow/structure)
	(Damp → Dry)
	• Temperature gradient (Hot → Cold)
	• Vapour pressure gradient
	(Moist → Dry)
	• Air pressure gradient (High → Low)
	• Reversion (steel → iron)
	(Processed → Original)

2 SEPARATE LIVES	• Differential movements
	• Differential durability
	• Incompatible materials
	• The process of assembly

Control

3 CREATIVE PESSIMISM   An allowance made for *Uncertainty* because the properties of all materials, their assembly and performance and the way a building is used, are neither totally ideal nor totally predictable.

Objectives

4 CONTINUITY	• Structure
	• Thermal insulation
	• Sound insulation
	• Fire protection
	• Cavity separator
	• Damp-/waterproofing

5 BALANCE	• With surroundings
	(temperature/moisture)
	• Laminates (plywood/facings)
	• Between the parts and the whole

Note: The relevance and importance of the principle of continuity is demonstrated vividly in the BRE Report, *Thermal insulation: avoiding risks*, 1989.

Rules and precautions
Examples include:
1 *Roof drainage*: For proper drainage of a flat roof, a fall of at least 1 in 80 is required. There is a further rule/ precaution to *compensate* for inaccuracies in building (if not Murphy's law): that the designed fall should be 1 in 50 (*Principles of modern building II*, 1961), but this has now been increased to 1 in 40 (PSA, *Technical Guides to flat roofing*, 1987, or Tarmac's *Flat roofing, a guide to good practice*, 1982) – the effects of 20 years' experience.
2 *Cavity wall width*: A cavity less than 50 mm wide is impracticable, so a minimum width of 50 mm is the rule.
3 *Flat roof cavity height*: For proper ventilation in a cold roof a minimum height of 50 mm is required. Other rules apply as to the maximum length of the cavity or the area of ventilation at opposite ends of the roof. The latter as a percentage of the roof area has been changed for roofs in sheltered areas (from 0.4 to 0.6 per cent).

3, 4 *Details cannot nowadays be left to be sorted out on site with any degree of confidence.*

2 Lessons for architects

2.01 Use of materials
Architects should understand the discipline that the combined use of materials, especially in modern multi-layer constructions, **1**, imposes on design solutions; or, more creatively, use this discipline as a motivation in design, **2**. In particular, the physical interactions involved need to be understood and allowance made for them.

Special care should be exercised in design when adding thermal insulation to 'traditional' forms of construction.

2.02 Determine the risks
They should recognise that modern building is not risk-free and that the best available techniques should be used to determine the risks in particular circumstances, **1**. These risks, together with the criteria on which they have been based, should be made known to (and understood by) their clients.

5 Mastic filling is not the all-purpose answer for inadequately designed joints.

2.03 Resist limitations imposed by cost
Architects should be realistic about the limitations that cost-limits or yardsticks impose on the standards that can be achieved, and resist the temptation of trying to get too much out of too little.

2.04 Reappraise life-span
The significance of the sixty-year economic life of buildings should be reappraised because, if taken too literally, buildings could be beyond economic repair after sixty years, if not before.

2.05 Check building regulations
It should be recognised that conformity with the requirements of building regulations and by-laws (or building control officers' interpretation of them) does not guarantee a successful building in all respects. (Conformity with the 1976 Building Regulations – and earlier – led to increased condensation risk because permanent ventilation was not required and because thermal insulation was in no way related to the usage of the building.)

2.06 Compensate for lack of craftsmen on site
Architects should recognise that the disappearance of the craftsmen on site, who used to be able to sort out detail difficulties, **3**, **4**, imposes on them the need for closer three-dimensional examination (including making models if necessary) of proposed details and the subsequent need to communicate the details chosen, clearly and explicitly, to those who will be responsible for their assembly on site. In addition to carefully thought-out and clearly drawn details, rigorous specification and site supervision are required to ensure that the designer's intentions are carried out. Site personnel should be given an unmistakable indication of the quality of execution which will be insisted upon.

2.07 Increase factors of safety
It should be understood that the understandable tendency to reduce, as much as possible, factors of safety without a sound quantitative (or sometimes reliable empiric) basis leaves the building with very little, if any, 'fat' with which to counteract unknowns – the technological balance is usually so fine that it does not need much to alter it, with deleterious results. It is almost true to say 'the better the science, the smaller the safety factor' – and the less leeway for faulty workmanship and abuse of the completed building.

2.08 Use of quantitative data
Architects should learn to use and interpret quantitative data, to understand that empiric and science-based approaches cannot be mixed (except with great insight), and to accept the need for a more rigorous engineering approach to those problems that used to be solved satisfactorily by empiric means (for example, fixings and wind loading especially, and provision for differential movements, both of which have structural implications).

2.09 Recognise contribution of specialist contractors
Architects should recognise that specialist contractors may not necessarily be knowledgeable or expert in all matters to which their work relates. The scope of their expertise should be checked and not assumed, in particular when such contractors are asked specifically to undertake a design function. How this function is to be handled contractually also needs to be carefully considered.

2.10 Recognise high cost of repairing modern constructions
Finally, they should recognise that modern types of construction are extremely costly to repair if failures occur, due partly to their greater rigidity, partly to the greater use of adhesives and partly to the greater use of sealants. As regards the latter, there is a strong case for architects to stop having such unparalleled faith in sealants to fill each and every gap in a building, **5**. They have their uses, but must be used appropriately, and it must be remembered that no sealants will last the life of a building. Joints are a matter of design.

Study 2
Investigation kits

1 *The author with parts of his basic kit that consists of the basic and specialised kits described in the text. Case, as shown, has rules, screwdrivers, sketch book, brushes, knives, scraper, magnetic and claw pick-up sticks. (The base houses electronic and other measuring instruments).*
(Back, from left to right) video cameras, compact cameras, drill, camera with Borescope probe attached.
(In front of case, from left to right) Nikon camera with lenses for it in front, mirror, electronic instruments for salt detection, moisture (with probes) and condensation and level indicator – the latter also in 6.
(Front row, from left to right) Polaroid Image camera, electronic caliper gauge (manual version in 2), rule and tape. (Photo: Ben Rice.)

1 Need to see hidden building parts

Conventional builders' tools and equipment for opening up apart, one or two kits may be required for an investigation. A general-purpose or *basic kit*, 1, is required in all investigation and would include equipment for recording data, making measurements (mainly of dimensions, levels, temperature and relative humidity), making observations (principally of surfaces and, to a limited extent, of hidden parts of a construction) and limited opening up of a construction. Some of the equipment included in this kit may be home-made. Nowadays there is a wide range of instruments and meters that are electronic. They are more expensive than their 'manual' or 'analogue' counterparts but have the advantage, apart from being more accurate, of

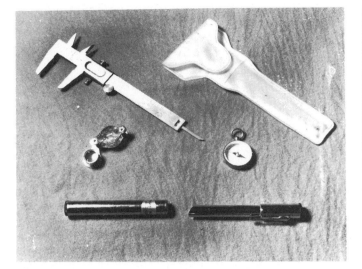

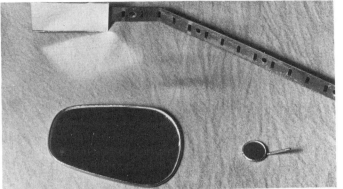

4 Mirrors mounted on handles, for inspecting hidden parts.

2 Anti-clockwise, from top left: callipers gauge; magnifying lens; pens; compass; callipers gauge case.

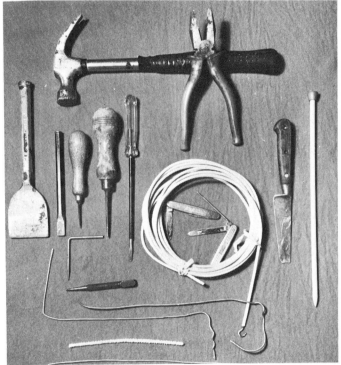

3 Small torch with two-way light source; ordinary torch (top fluorescent); large torch with quartz bulb (1000 m) beam; and car reading light attached to video battery (gives bright shadowless light).

5 Collection of conventional and unconventional tools for exploring and opening up operations.

being easier to read. Reference is made to both types. A *specialised kit* would be required where more accurate measurements and more extensive opening up for observation and/or laboratory testing is required. Most of the equipment in this kit is specialised (and usually expensive) and needs to be used by trained technicians.

Brief notes on the use of or need for each item of equipment are given below for each kit.

1.01 Basic kit

- Sketch pad, notebook, clipboard or small portable drawing board, scale, coloured pens and/or pencils, chalk and eraser. For recording all data in sketch and written form.
- Rules, tapes and small callipers gauge.
  Rules and tapes for measuring overall dimensions and profiles of details should be available. A small (say, 150 mm) relatively inexpensive callipers gauge is most useful for measuring widths of joints or cracks, overall

dimensions of frames, pipes, etc. and depths, of hidden parts such as joints especially, **2**. (An electronic calipers gauge, although more expensive and more accurate, is easier to read.) Use a micrometer if the exact thickness of a material is required (usually only necessary with thin metal sections).

- Adhesive materials for temporary fixing of rules, labels, etc. Blutak can be most useful and does not cause any damage.
- Cameras – 35 mm, polaroid and video. (Fast almost grainless colour film and low light video cameras can give good shadowless results, making flash and video lights virtually unnecessary).

A photographic record is an invaluable *aide-mémoire*. Subsequent study of the photographs or videorecording often reveals important 'clues' missed during the actual inspection or opening up. The value of any photographic technique can be enhanced if suitable labels are affixed near the object being photographed.

Colour photographs are now the norm – they are cheaper than black and white; developing and printing of colour film can be obtained quickly almost everywhere.

Colour slides have the advantage that they can be viewed by back projection without darkening the room. The back projector's screen can be used as a light box for tracing off details or marking important aspects. The newer polaroid photographs are useful on site for writing on – they need a felt pen that will write on glossy/plastic surfaces.

A video camera has the advantage of including sound and can replace a note-taking tape recorder. It can cover much ground and there is less chance of something or some part being missed. Moving objects (e.g. dripping water and pulling apart in opening up) are illustrated vividly. Colour comes as standard but pictures can be viewed in black and white simply by twiddling a knob on the television. Slowing down or stopping a recording is invaluable, but particular parts of a recording cannot be located quickly. Reasonably good quality off-screen photographs are expensive. Edited video recordings are invaluable to guide laymen, such as clients, lawyers and judges (the Official Referee's Court has facilities for showing video recordings).

- Compass
- Torch or other light source, **3**. Ordinary torches have the disadvantage of producing a light with shadows. This can be distracting. Light sources with quartz lamps (for example, video lights, car reading lights that can be plugged into a video battery or the light source from an inspection kits for cavities) are shadow free.
- Mirrors.
  Mirrors, **4**, can be very useful for visual inspection of otherwise inaccessible exposed parts of a construction, such as the underside of parapet coping projections, and tops of flashings; or of voids such as floor and roof spaces in a construction after limited opening up. In the latter instance the use of a mirror with an appropriate light source restricts the amount of opening up required for the inspection. Wide-angle mirrors (car wing mirrors, for example) greatly extend the area that can be viewed at one time. Other 'refinements' that can extend the usefulness of mirrors include mounting them on handles or rods and at different angles. Whatever mirror is used, care is required in the interpretation of the mirror image. Still less opening up is required if a Borescope is used; see 1.02 'Specialised kit', 'Visual inspection probes' later.
- Magnifying glass, **9** (triple lens), pocket microscope.
  Helpful in identification of the condition of surface finishes, debris in cracks, nature of holes, and fungi. What is seen also needs careful interpretation.
- Binoculars.
  Indispensable for studying surface defects and details that are inaccessible without the use of ladders or scaffolding.
- Penknife, bradawl, screwdriver, small hammer, pliers and other 'tools' for exploring and excavating on a limited scale, **5**. These items, and others more unconventional such as skewers, pieces of wire and wire hangers, are useful for scraping surface finishes and removal of debris, or limited opening up of parts of the construction.
- Small spirit level or level and angle indicator, **6**. Some small tapes come with a spirit level. Angles and grades can also be measured electronically.
- Moisture meter, **7**. The one illustrated, with prongs, has an analogue scale for indicating the 'scale' of dampness. Newer versions incorporate flashing lights that are easier

6 *Level and angle indicator (top); spirit level (bottom).*

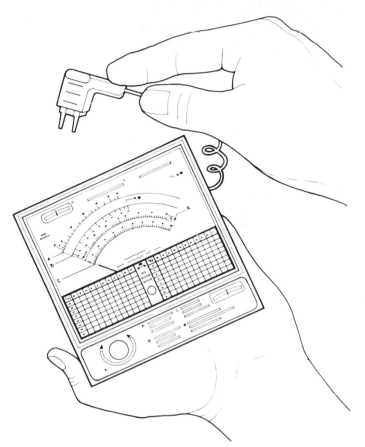

7 *Moisture meter: prongs are inserted into building fabric and moisture content read off.*

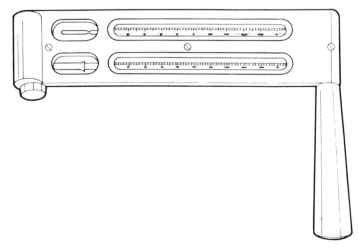

8 *Whirling hygrometer with wet and dry bulb thermometers, for determination of relative humidity.*

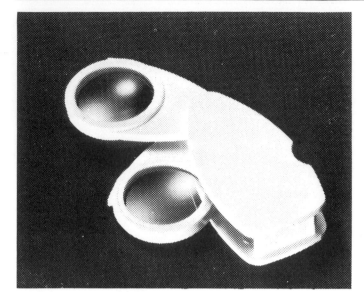

9 *Multi-lens magnifying glass*

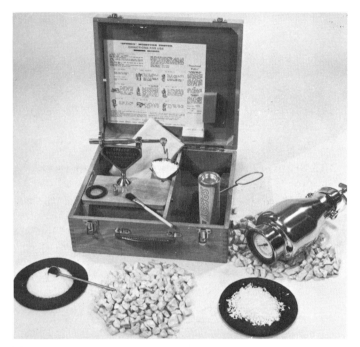

11 *'Speedy' moisture meter.*

10 *Taking a core sample from a reinforced concrete basement wall.*

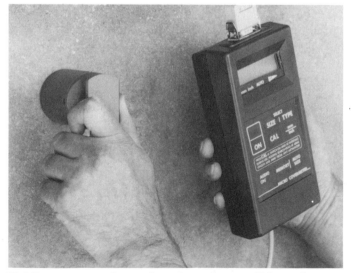

12 *Micro cover meter for non-destructive gudiance to thickness of concrete cover to steel reinforcement (Courtesy of Kolectric Ltd, Thame).*

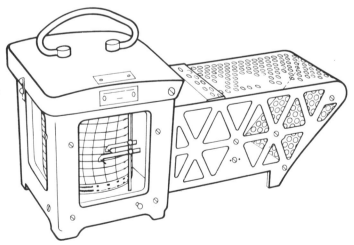

13 *Temperature and humidity recorder, inscribing continuous record of both variables (in two colours) on graph paper-covered drum at left.*

to read. Other moisture meters have pads that are placed on the surface. Some of these are reputed to measure the average depth of dampness. All meters are useful for obtaining a crude assessment of dampness of most material except timber, for which reading from most meters are usually reasonably accurate, those with prongs in particular. All meters should be used with caution and their limitations understood.

- Whirling hygrometer, **8**, or an electronic version that is easier to read.
- Thermometers for measuring air or surface temperature. Electronic versions are now available and some are capable of providing continuous recordings.

Useful for basic guidance only, if condensation is the likely cause of dampness.

1.02 Specialised kit

- Power drill.

For taking dust samples for moisture determination or for taking small cores of plaster or brick for inspection (see also 'Speedy' moisture meter later).

- Diamond/carborundum sawing and drilling.

For opening up or taking cores, **10**. Requires specialist to operate and may require the use of water. *Cores* can be useful for moisture determination but are particularly helpful if there is a need to identify the layers of materials actually used in the thickness of a construction – for example, whether a dpm was actually laid in a ground floor or whether a vertical dpc was actually incorporated into a wall – without extensive opening up. For details of techniques, including thermic lacing, see 'Cutting a way through problem materials' by Paul Phillips and David Lazenby, *AJ* 22.10.75, pp. 855–866, CI/SfB (D4). Although intended for refurbishment and alteration work, the techniques described in this article are equally suited to diagnostic work.

- 'Speedy' moisture meter, **11**.

Gives far more accurate readings of moisture content from dust samples than moisture meters.

For moisture determination from dust samples and cores see **6.03** under 'Tests/measurements', in Study 3.

- Visual inspection probes.

Sophisticated optical instruments with an integral light source using fibre optics are available for inspecting inside cavities and voids without extensive opening up – a small-diameter hole to gain access to the cavity or void is all that is required. These instruments are available with different angles of view and take some getting used to. Special attachments for a camera are available. (Olympus Industrial Borescope made by KeyMed, Southend-on-Sea, Essex, comes in a wide range of angles of view, lengths and diameters. Many attachments are also available. Private investigators and firms such as Laing or Wimpey undertake surveys of cavities – see list of testing laboratories.)

At a less sophisticated level, a *modelscope* can sometimes be used to advantage provided it is possible to devise an appropriate light source. It has a wide angle of view.

- Cover meter, **12**.

Useful for giving guidance on cover to steel reinforcement in concrete by non-destructive means.

- Temperature and relative humidity recorders. Electronic recorders are now available for measuring dry and/or wet bulb air and surface temperatures. Most are battery operated and can record continuously over a given period. Temperature and relative humidity recorder (an older version in **13**) are essential for obtaining appropriate recording in condensation problems.

- Infrared thermometers for surface temperatures. Portable electronic versions are available and their use can be helpful in some situations. If a comprehensive survey of the variation of surface temperature of a large area is required, an infrared recording by a specialist laboratory is essential.

1.03 Selective list of research and development bodies

Research and development serving the construction is carried out by:

(a) Government research establishments of which the BRE is the main organisation.
(b) Research and information associations
(c) University and polytechnic departments.
(d) Research and development in contracting and manufacturing firms.

The list below is a selection of the above. Some of the organisations may also carry out tests on materials – see **1.04** later.

Government/Local Authority research establishment
BRE (Building Research Establishment), Department of the Environment.

(1) Building Research Station (including Princes Risborough Laboratories), Garston, Watford, Hertfordshire WD2 2JR Tel: (0923) 674040
(2) Fire Research Station, Boreham Wood, Hertfordshire WD6 2BL. 081-953 6177
(3) BRE Scottish Laboratory, Kelvin Road, East Kilbride, Glasgow G75 0RZ (03552) 33001.

Metereological Office, London Road, Bracknell, Berkshire RG12 2SZ (0344) 420242
LSS (London Scientific Services) (GLC Scientific Services Branch reconstituted to serve the London boroughs and others), Room 755MB, The County Hall, London SE1 7PB 01-633 5975. *Note*: Check address and telephone as the service will relocate after County Hall is sold.

Research and information services
BCA (British Cement Association), successor to C&CA, Wexham Springs, Slough, Berks SL3 6PL Tel: (0281) 62737 Fax: 0281 62251/63727/63851
BCRA (British Ceramic Research Association), Queens Road, Penkhull, Stoke-on-Trent, Staffordshire ST4 7LQ Tel: (0782) 45431
BCSA (British Constructional Steelwork Association Ltd), 4 Whitehall Court, London SW1A 2ES Tel: 071-834 1713 Fax: 071-976 1634
BDA (Brick Development Association), Woodside House, Winkfield, Berkshire SL4 2DX Tel: (0344) 885651 Fax: (0344) 885651 (Ext 201)
BRMCA (British Ready Mixed Concrete Association), 1 Bramber Court, Bramber Road, London W14 9PW Tel: 071-381 6582
BSRIA (Building Services Research and Information Association), Old Bracknell Lane, Bracknell, Berkshire RG12 4AH Tel: (0344) 426511
CDA (Copper Development Association), Orchard House, Mutton Lane, Potters Bar, Herts EN6 3AP Tel: (0707) 50711 Fax: (0707) 42769
CONSTRADO (Constructional Steel Research and Development Organisation), NLA Tower, Addiscombe Road, Croydon CR3 3JH Tel: 081-688 2688

FRCAB (Felt Roofing Contractors' Advisory Board), Maxwelton House, Boltro Road, Haywards Heath, Sussex RH16 1BJ Tel: (0444) 51835

LDA (Lead Development Association), 34 Berkeley Square, London W1X 6AJ Tel: 071-499-8422

MATAC (Mastic Asphalt Technical Advisory Centre), Lesley House, 6–8 The Broadway, Bexleyheath, Kent DA6 7LE Tel: 081-298 0414 Fax: 081-298 0387

PRA (Paint Research Association), 5 Waldegrave Road, Teddington, Middlesex TW11 8LD Tel: 081-997 4427 Fax: 081-943 4705

RAPRA (Rubber and Plastics Research Association), Shawbury, Shrewsbury, Shropshire SY4 4NR Tel: (0939) 250383

TRADA (Timber Research and Development Association), Stocking Lane, Hughenden Valley, High Wycombe, Buckinghamshire HP14 2ND Tel: (0240) 243091 Fax: (0240) 245487

Water Research Centre, Medmenham Laboratory, Henley Road, Medmenham, PO Box 16, Marlow, Buckinghamshire SL7 2HD (0491) 66282 Fax: (0491) 589094

ZDA (Zinc Development Association), 34 Berkeley Square, London W1X 6AJ Tel: 071-499 8422

1.04 Selective list of testing services

The laboratories listed below carry out a major part of their testing on materials and products in the construction industry. Check first as to the test they do beforehand. All are NATLAS testing laboratories.

(NATLAS – National Testing Laboratory Accreditation Scheme, set up by the Department of Trade and Industry at the National Physical Laboratory, Teddington, Middlesex TW11 0LW, Tel: 081-977 3222.)

An asterisk (*) after the name of a testing laboratory indicates that the range of services offered may be limited.

Bedfordshire County Council,* Engineering Laboratory, Austin Canons Depot, Bedford Road, Kempston, Bedfordshire MK42 8AA (0234) 45493

British Ceramic Research Association Ltd.,* Queens Road, Penkhull, Stoke-on-Trent, Staffordshire ST4 7LQ (0782) 45431

British Gypsum Ltd,* Research & Development, East Leake, Loughborough, Leicestershire LE2 6JQ (0602) 214321

British Standards Institution,* Test House, Marylands Avenue, Hemel Hempstead HP2 2SQ (0442) 3111

The Chatsfield Applied Research Laboratories Ltd,* 13 Stafford Road, Croydon, Surrey CR0 4NG Tel: 081-688 5689

Fairclough Civil Engineering Ltd.,* Geotechnical Services (Northern Div.), Chapel Street, Adlington, Lancashire PR7 4JP (0257) 480264

Hampshire County Public Analyst and Scientific Adviser,* Hyde Park Road, Southsea, Hampshire PO5 3LL (0705) 828965

Sir Alfred MacAlpine & Son (Southern) Ltd,* Kingswood Laboratory, Holyhead Road, Albrighton, Wolverhampton, Staffordshire WV7 3AR (090 722) 2455/6/7

Sir Robert MacAlpine & Sons Ltd,* The Laboratory, Rigby Lane, Hayes, Middlesex UB3 1EU Tel: 081-573 2637

Laing Technology Group Ltd, Page Street, London NW7 2ER Tel: 081-959 3636/Fax: 081-906 5297

Sandbergs, 40 Grosvenor Gardens, London SW1 0LB. Tel: 071-730 3461/Fax: 071-730 4972

Harry Stanger Ltd, The Laboratories, Fortune Lane, Elstree, Hertfordshire WD6 3HQ Tel: 081-207 3191/Fax: 081-207 5878

Wimpey Laboratories Ltd., Beaconsfield Road, Hayes, Middlesex UB4 0LS Tel: 081-573 7744/Fax: 081-848 1554

UK Analytical Limited, 15 Chatsworth Terrace, Harrogate, North Yorkshire HG1 5HT Tel: 0423 522005/Fax: 0423 523285

Yarsley Technical Centre, Trowers Way, Redhill, Surrey RH1 2JN (0737) 65070/9

Study 3
Diagnosis: principles and procedures

1 Diagnosis of defects

1.01 Impartiality and realism

The diagnosis of a defect should represent the impartial assessment of all the data available and should not be (as Eldridge puts it) 'a means of confirming an opinion already formed'. In simple terms, the causes and related factors that have been responsible for a defect are determined by relating the answers from many and varied searching questions to the known behaviour of the materials involved. Consequently, *the quality of the questions asked and the answers to them are important determinants*. The aim of this section is to describe or explain the essential requirements and scope of the investigation that needs to be carried out in making the assessment of the data collected. Guidance has already been given on the kits that should be useful for opening up and tests.

2 Basic requirements

2.01 Need for system and method

The investigation required for the diagnosis of the cause(s) responsible for a building defect has to be carried out thoroughly and systematically. The proper identification of the cause(s) is essential if appropriate remedial work is to be devised and if, as is so often required, liability for the cost of such work is to be fairly established and agreed. Although the question of liability should not influence the deduction of the cause(s) of a defect, it could (and often does) determine the scale of the investigation that may be required.

Apart from having a great deal of patience (among other things the temptation to jump to conclusions too soon has to be strongly resisted), inquisitiveness and inventiveness, the investigator must have a good knowledge not only of building construction and the building process but also of the characteristics and behaviour of materials in construction and the factors likely to influence their performance in use. The investigator has to learn to observe comprehensively; to understand what he may see with visual aids (e.g. in mirrors, through magnifying glasses, probes and binoculars); to devise techniques and procedures whereby valuable evidence is not destroyed during the process of opening up a construction; and to assess very carefully the results of non-destructive tests.

3 Basic procedure

3.01 Forensic approach

Crime detection, pathology or even forensic medicine provide the precedents for the procedure that should be followed in the diagnosis of building defects. Clues found or sought at different times during an investigation have to be analysed and false trails identified. Background information has to be gathered, analysed and its relevance assessed;

specific evidence has to be found (usually, but not always, by opening up a construction or by other destructive means), inspected, analysed and assessed; hypotheses have to be tested and comparisons made. As with design, a simple straightforward linear procedure is seldom possible. The process is inevitably an iterative (cyclical) one and the more unusual, unfamiliar or complex the cause – and more than one cause may be responsible for the defect – the more the need to go back and repeat some or all of the stages previously completed.

3.02 Decide scope; gather evidence

The basic procedure starts with the way the cause of the defects has manifested itself (the *symptoms*, medically). The symptoms provide the first clues, **1–3**, and these enable the

1 *The 'symptom' (visible manifestation of failure) provides the first clue to cause of defect, and suggests scope and nature of investigation. In this case the diffuseness of the damp (as well-defined patches or bands), and the presence of mould (which requires relative humidity of air to be above 70 per cent to allow growth) suggest condensation as probable major cause. Investigation must be designed to test this tentative deduction. See also **2** and **3** overleaf.*

2 *Suggests rising damp (damp does not extend beyond lower part of wall and ends at characteristic tide mark);* **3** *suggests penetrating damp through localised perforations.*

basic *scope* of the investigation to be determined, though other factors may also have to be taken into account: see 'Scope of an investigation' later. Then follows the investigation that includes the need for the collection and recording (often of a great deal) of data from a wide variety of sources (see 'Information required and its sources' later); the elimination of red herrings and misconceptions and (the most important and most difficult part) the comparison of the symptom(s) with the *known* behaviour of the materials involved and the conditions to which they have been exposed and under which they have had to perform. It is from the latter that the cause of the defect is deduced.

4 Causes/sources of defects

4.01 Multiple factors
One of the most difficult aspects of any diagnosis is that very often more than one cause may be responsible for the defect, although in most cases it is possible (and indeed necessary) to identify the primary cause. For example, movement may have initially been responsible for the cracking of part of an element. The crack(s) would then make rain penetration possible and the water penetration would manifest itself as dampness. In this case there would have been no dampness if there had been no cracking. But equally important, the existence of a crack and the appearance of dampness does not *necessarily* mean that the crack has provided the means of ingress for the water. In some cases the two defects are not related, although both will have to be remedied. Nothing can be taken for granted.

4.02 Three basic 'causes'
It is important at the outset to distinguish clearly between the *cause(s)* of a defect, and the agency or factor (*'source'* may be more descriptive) that has, so to speak, activated the cause(s). The cause(s) of a defect may be identified fairly easily as there are really only three basic ones, namely:

- dampness
- movements (i.e. physical change)
- chemical/biological change.

4.03 Many specific 'sources'
It is the 'sources' of these causes that can present problems. There are many of them and some are interactive. However, in any diagnosis they have to be eliminated systematically until the chief 'culprits' have been identified. It is the 'culprits' that have to be dealt with in any remedial work.

The possible 'sources' of each of the three causes can be summarised as:

- dampness
  1 rain
  2 ground
  3 construction process
  4 atmosphere (condensation notable)
  5 water supply
  6 faulty services
  7 maintenance and general usage
- movements
  8 externally applied loads (structural loading and movements in soils)
  9 changes in temperature
  10 changes in moisture (some of the sources for dampness relevant, but the atmosphere notably)
  11 vibrations
  12 other physical changes (such as ice or crystalline salt formation; loss of volatiles, as in asphalt and sealants)
  13 chemical changes (most of the sources for dampness are relevant; corrosion, sulphate attack and carbonation are most important changes)
- chemical/biological changes
  14 dampness (corrosion, sulphate attack, wood decay)
  15 temperature (wood burning)
  16 solar radiation (fading and/or decomposition of paints, plastics, sealants)
  17 presence of incompatible substances (setting of cement, adhesives and sealants).

4.04 Interrelationship
In the final analysis, the extent to which any of these sources may influence the severity of a defect depends on the interrelationship between many things, including the nature, behaviour and transfer of the sources themselves; and the nature, behaviour and resistance to the various sources of materials (in combination especially). These aspects will tax the diagnostician.

5 Scope of an investigation

5.01
The scope of an investigation into the cause(s) of a defect is dependent primarily on the amount of money that can be spent on it (and investigations can be very costly). That, in turn, must be related to:

- *the nature* of the defect
- the *accuracy* with which the cause or causes need to be identified
- the main *reason* for wanting to know why the defect has occurred (i.e. to determine the remedial work required; or to ascertain liability).

Generally, the more uncommon the defect and the greater the accuracy required in the identification of the cause(s), the wider the scope of the investigation. At a very simple level, the diagnosis may be based on an investigation that consists of no more than a thorough visual inspection and limited data collection. At the other extreme it may be necessary to undertake extensive opening up, site and laboratory tests and intensive data collection: this could take months rather than weeks. Inevitably, each case has to be considered on its merits.

6 Information required and its sources

6.01 Intended result versus actual result

By definition, failure means something has not been successful. In other words, a difference (sometimes a considerable difference) has occurred between what was *intended* or expected and what has *actually* been achieved or experienced. The information required in diagnostic work should aim, first and foremost, to provide the investigator with data concerning the *actual* materials and details that were executed during construction and the *actual* conditions to which the materials and the elements of which they form part have been exposed or subjected before, during and after construction, **4**, **5**.

6.02 Aspects to be investigated

The sources of information from which the data have to be collected can be wide and varied (recorded, oral, from observation, published, and from tests). Whatever the source, data must be properly recorded. Before discussing these, it is appropriate to summarise some of the more important *design*, *construction* and *usage* aspects that could be inherently responsible for a defect either separately or in combination (i.e. one or more). These are:

- incorrect materials were specified and/or inappropriately detailed
- correct materials were specified, but inappropriately detailed
- correct materials were specified and appropriately detailed, but incorrectly ordered, handled and/or assembled on site
- inappropriate design conditions of loading or exposure used in making calculations or predictions of performance.

Note that the design conditions assumed may well accord with generally accepted good practice, but in the context of a failure the *actual* conditions are relevant, and it is in this respect that the design conditions used can be said to be

4a, b, c *Photographic record of actual conditions, using* in situ *notes and scales to aid interpretation.*

5 *Trying to dig out very damp flexcell from an expansion joint.*

inappropriate. Determination of the actual conditions and their possible influence in relation to a defect requires somewhat different and more accurate predictive techniques than those normally considered adequate for design purposes. In the absence of such techniques it is necessary to 'stretch' available design predictive techniques. For example, if interstitial condensation is suspected, the risk of such condensation occurring would be tested for a range of environmental conditions (temperature and relative humidity) both inside and outside, including extremes and for a range of vapour resistances of the materials used in the construction. The answers may not be all that accurate, but very often it is sufficient for the investigator to know what ballpark he is playing in.

6.03 Sources of information
Summarised, the main sources of information and the data they are likely to provide are as follows.

Drawings and spec
All the drawings and specifications (i.e. including those of consultants, specialists and sub-contractors) used during construction should theoretically provide the data required on the materials used and details of the construction. In practice, they cannot normally be completely relied upon to give these data as, apart from the poor or indifferent workmanship that may have occurred on site, they do not necessarily include all the revisions made during construction. Unless the accuracy of drawings and specifications can be verified, the information they contain can only provide basic clues – the construction actually executed will have to be determined by observation and inspection.

Architect's instructions
In the absence of up-to-date drawings and specifications, architect's instructions may provide information on modifications made during construction. All relevant instructions need to be examined carefully and compared to determine 'the state of play' at any given time. It is not uncommon for instructions on the same construction or material to be varied often during the course of a contract.

Site notes, minutes, reports
Apart from providing information on modifications made during construction, these, taken together, should (though they do not always) provide a clear picture of:

● the sort of difficulties encountered during construction (e.g. materials supply, labour supply, site conditions, problems of assembly)
● the quality of workmanship achieved
● the extent to which precautions were taken to protect materials stored on site, the elements during construction and the building as a whole from the weather.

Note that knowledge of the degree of exposure to the weather during construction could be very helpful in the diagnosis of defects where dampness has been identified as the cause, particularly when there is little, if any, evidence of a source of water (e.g. rainwater, groundwater or condensation) being responsible since completion of the building. If necessary, consult the Met Office – see 'Meteorological records' below.

● Weather conditions experienced during construction.

Note that knowledge of these conditions, which ideally should include rain, wind, sun and snow, could be helpful in the diagnosis of a wide variety of defects. During construction, the weather could influence the amount of rainwater absorbed by materials and elements (the degree of protection mentioned earlier would have to be taken into account); the amount and rate of drying out that may have taken place and movements. If necessary, consult the Met Office – see 'Meteorological records' below.

Maintenance manuals and records
If in existence, the manual should contain reasonably accurate details of the construction and the record of details of alterations and additions carried out since the building was completed. As important, the record should give details of materials used, in cleaning and redecoration especially, and the history of any defects.

Reports on the defects
Any existing reports on the defects may contain information and opinions that should be considered in the light of further evidence available.

Interviews
Interviews with any of those connected with the design, construction, maintenance and use of the building can provide information on a number of aspects associated with a defect, the most important being its history and (from the users of the building) some indication of conditions of exposure. Information obtained from interviews needs to be treated with some caution. It may not always be entirely accurate (the way questions are asked may influence this) but, even so, could provide some useful clues to be followed up.

Meteorological records
These may have to be referred to for weather conditions during and after construction. The Meteorological Office (local station or Bracknell – see Study 2, 1.03) can supply a wide range of measurements for many stations in the United Kingdom. Duration and frequency of rainfall (usually at hourly intervals) can be most useful. (Site records usually provide a crude measure of rainfall, with weekends generally excluded.) Hourly values of temperature, relative humidity, wind and sunshine are also available. DRIs for specific locations are available on request. Meteorological data needs careful interpretation; the relationship between the Met. Station and a particular site is important.

Inspection
Inspection of the defect and, as important, the context in which it has occurred provides the most essential information and the clues that have to be followed up. The investigator requires a very keen sense of observation, making use not only of sight but also, as occasion demands, of hearing, smell or touch. Site inspections cannot be hurried and often it is better to carry out the inspection in stages. Time is needed for the observations to be mentally digested and for comparisons to be made with other data already collected. Such comparisons may indicate the need for the collection of more data, the scope of any further inspections, opening up of the structure or the need for specific tests.

Published information/research
A need is most likely to arise for detailed information on specific aspects such as the properties of materials, test procedures and the anticipated performance of materials in given situations and conditions. Very useful information

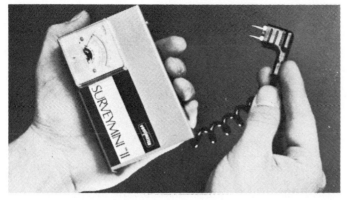

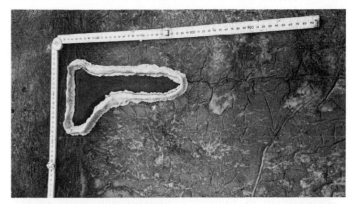

6 *Electronic moisture meter which will assist in locating areas of dampness. Samples of damp material can then be sent to laboratory for accurate analysis.*

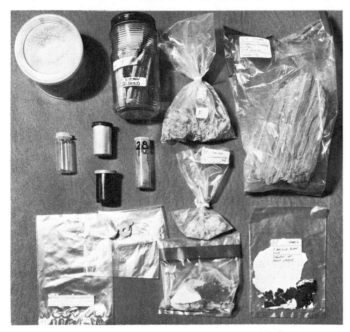

7 *Dust and other samples of materials and cores stored in bottles or plastic bags as records of evidence uncovered during investigation, or for analysis.*

8a, b, c *Using dams to test the possible location of leaks in flat roofs, prior to embarking on an opening-up exercise.*

can be obtained from research papers published by BRE and other research or trade associations, such as BCA, BDA, TRADA. In some very difficult cases it may be worth finding out whether any of the universities or polytechnics are carrying out research in the area of interest.

British Standards and Codes of Practice will have to be consulted constantly.

Finding information that was current when the building was either designed or constructed – and it is essential to use contemporaneous information, in litigation especially – can present difficulties.

Tests/measurements (see Study 2, 1.04, for list of testing laboratories)

Tests and measurements of properties of materials and of conditions may include:

● Moisture determination
For timber, a portable electrical resistance moisture meter provides sufficiently accurate results, **6**.
For porous materials such as plaster, brick and concrete blocks, portable electrical resistance moisture meters provide general guidance only. More accurate results are obtained from measurement of dust samples either chemically on site (using the 'Speedy' moisture meter) or

by oven drying in a laboratory. Guidance on the effect of deliquescent salts may also be obtained by exposing dust samples to a constant relative humidity of 75 per cent in a laboratory (usually by using a salt solution in a desiccator). Cores of about 25 mm diameter may also be used for moisture determination (including moisture determination in solid concrete walls and slabs) by oven drying in a laboratory. An advantage of using either dust samples or cores, **7**, is the ability to determine moisture gradients or profiles within the thickness of a construction.

For details of techniques see:
1 BRE Digest 245, *Rising damp in walls: diagnosis and treatment.* (This digest now incorporates BRAS Information TIL29, *Diagnosis of rising damp,* April 1975.) Deals with drilling technique and describes use and limitations of 'Speedy' moisture meter, measurement of hygroscopicity (using a desiccator) and moisture content (using oven drying), and gives guidance on the interpretation of results.
2 Newman, A. J., *The independent core method – a new technique for the determination of moisture content,* BRE Current Paper CP 7/75. January 1975. Describes technique for taking cores from brickwork, but principles are equally applicable to concrete.

● Chemical analyses
Few, if any, chemical analyses of the composition of materials can be carried out on site (ultrasonic techniques apart). Appropriate samples have to be taken from the building and sent to a laboratory (either a commercial laboratory, or some university, polytechnic or technical college departments may be prepared to undertake an analysis). It is important to establish first whether or not a chemical analysis is going to be of any value. For example, bricks contain a good deal of silicon, so any analysis to determine whether a silicone-based dpc has been injected is not going to be of much use. It may sometimes be necessary to know the chemical composition of rainwater, underground water or the atmosphere. Samples taken from the site can be analysed by a laboratory. The local public analyst may be able to help (a charge is made if he carries out tests).

● Physical analyses
These may be related to structural properties, ability to absorb and transmit water, resistance to frost, adhesive qualities and range of moisture or thermal movements. Again, tests have to be carried out in a laboratory (commercial or educational institutions – see 'Chemical analyses' above). Such tests may be comparative, i.e. made on materials similar to those used in the building under investigation.

Measurements that can be made on site include:

1 *Temperature* of air or surfaces and the relative humidity of the air. To be useful, such measurements have to be made over a period of time using suitable recording apparatus with appropriate probes. (Some university, polytechnic or technical college departments may be prepared to loan the equipment which needs to be properly calibrated and operated.)

2 *Surface temperature*: A comprehensive survey of the temperature of surfaces is best carried out using infrared recording by a specialist laboratory.

3 *Rate of ventilation* using tracer gas techniques or sensitive (electronic) anemometers. (Specialist equipment requiring specialist operation.)

4 *Movement of cracks*. The traditional glass telltale has limited use – if the glass cracks there has been movement but the amount and direction of this movement cannot be established. More useful are pins, discs or other marking devices, including transducers fixed on either side of the crack so that the distance between the pins or markers can be measured from time to time. See:

  1 BRE Digest 343 *Simple measuring and monitoring of movement in low-rise buildings,* Part 1: cracks, April 1989.

  2 BRE Digest 344, *Simple measuring and monitoring of movement in low-rise buildings,* Part 2: settlement, heave and out-of-plumb, May 1989.

Note Telltales in acrylic are now commercially available for a range of locations (Avonguard Products Ltd, 61 Down Road, Portishead, Avon BS20 8RB – 0272 849782).

Simulation

It is sometimes necessary, if not essential, to simulate sources of dampness such as rain. The rigs or techniques that may be appropriate usually have to be determined experimentally or laboratory type rigs transferred to site. Sprays for walls or dams for roofs are most commonly used. Dam tests that use dyes, ordinary or fluorescein, are often essential to determine location(s) of leaks on flat roofs. Tests of the outlets and of the roofing generally are normally required. A detailed description of the relevant tests is given in PSA, *Flat roofs, Technical Guide* (2nd edition), Annex E. (Using fluorescein in dyes with an ultraviolet light is not included. These dyes need to be used with care, the interpretation of what is seen especially.) At a more sophisticated level, tests of the watertightness of windows to BS 4315: Part 1: 1968: *Windows and structural gasket glazing systems* can be carried out on site.

Opening up

Opening up (small or limited opening up apart) is usually carried out by a builder or craftsmen, often using specialist equipment, **9, 10** (see Study 2, 1.02, 'Diamond/ carborundum sawing and drilling').

9, 10 *Using carborundum saw for,* **9,** *cutting through reinforced concrete loading slab;* **10,** *cutting through the steel reinforcement.*

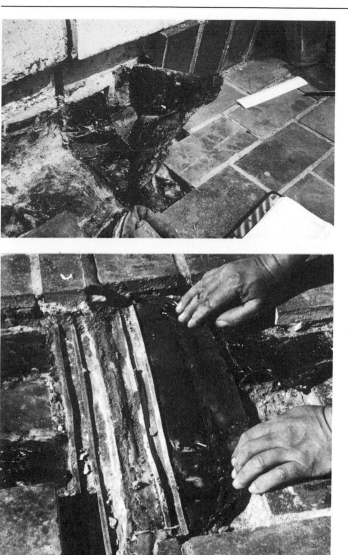

11, 12, 13 *Opening up: details dissected that range from the complex (11), to the curious (12), to the surprising (13).*

The nature and scale of opening up depend on circumstances, but in all cases it must be carried out by skilled workmen (better, craftsmen if they are available) who must be clearly briefed on the scope of the work, the need for care and protection and the need to preserve evidence as intact as is humanly possible, **11–13**. Importantly, all concerned should recognise that what is required is *not demolition work* but surgery tempered with patience.

7 Analysis and communication

7.01 Needs
Having gathered together all the evidence available, the investigator then has to analyse it to determine what are or are likely to have been the main and/or contributory cause(s) of the defects. Having made his or her analysis, the next (and, in some ways, the most important, if not difficult) job is to communicate findings to others. And the others are invariably laymen, from the client to insurers, lawyers or judges.

7.02 Simplicity
This is not the place to discuss or give advice on writing a technical or an expert's report. There are a number of good books on the subject, and a selective list is given at the end of this study. Suffice it to say that most of the readers of reports on investigations of building failures do not understand the technicalities nor the technical terms associated with buildings and their constructions. They need to have everything explained in the simplest terms that they can be expected to understand. For the sake of any other technical person or expert reading the report, the technical term or description expressed colloquially can, with benefit, be given in parentheses in the appropriate places.

7.03 Use of overlays, coloured pens and principles
Overlays of either tracing paper or acetate sheets are invaluable for analytical drawing over details from architect's drawings (or other sources) without defacing the latter. Acetate overlays with opaque drawings can be used with a copier and save much time. Pens of different colours can, for example, be used to advantage to track the likely path(s) of water in and out of a part or parts of a construction, or the direction of the flow of heat and water vapour or of thermal or moisture movements. By applying the five principles described earlier in Study 1, 1.03, page 3, the mechanics of what might have gone wrong or the factors that were likely to be have been influential (i.e. the nature of the problem) might be better understood – and without the aid of a computer. (The latter can, of course, be brought in once the nature of the problem has been defined.)

7.04 Three-dimensional techniques
It almost goes without saying that visual material by way of annotated photographs, sketches or models (or, to be really up to date, audiovisually by way of videorecordings) are an invaluable aid to unravel complexity. The investigator may find that in many cases there is a need for sketches and/or models (of a simple kind) to try to understand complex details or how water, for example, may track in or out of parts of a construction. Figure **14** is an example of a three-dimensional sketch of a jamb with two different positions for the window frame at different levels of the same elevation. The sketch was prepared to illustrate principles to the client and his lawyers. Figure **15** is an example of the same condition as **14(a)** but so drawn to analyse the complex paths of the penetrating water that eventually caused leaks. Having done its analytical job it was then used to explain the problem to the client and his advisers.

Selective references for report writing
1 Mildred, R. H., *The expert witness*, George Godwin, London and New York, 1982. Primarily intended to guide the expert witness through all stages of litigation.

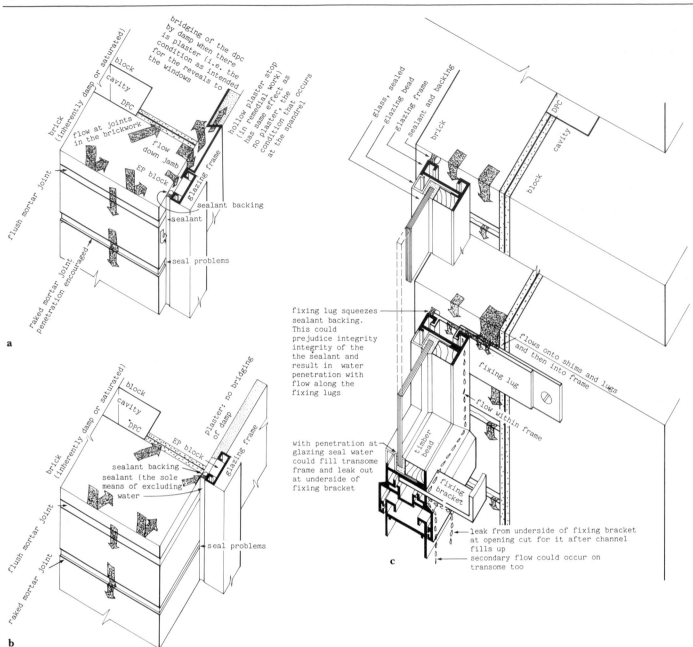

14 *Three-dimensional analytical sketches of window jamb showing possible directions of flow.* (**a**) *with glazing frame in front of dpc and* (**b**) *behind dpc.*

15 *Three-dimensional analytical sketch of flow and leakage of condition shown in* **14(a)** *and looking from inside.*

Does so in a manner fatherly, succinct and clear. Similarly with brief and to the point guidance on an expert's report – Chapter 5, page 21.

2 Chappell, David, *Report writing for architects*, Butterworth Architecture, London, 1984. Deals mainly with reports an architect would normally prepare. Comprehensive and well presented with helpful guidance on structure and format. Reports on investigation into defects are covered in 5.4, pp. 73–80. Chapter 7 has advice on proof of evidence.

3 Reynolds, Michael P. and King Philip S., *The expert witness and his evidence*. More formal but more comprehensive than Mildred. Includes advice on all types of documents needed in litigation.

4 Hamilton, Alaine, *Writing matters*, RIBA Publications, 1988. Beautifully and well written by someone who knows the subject and the needs of architects including an excellent chapter on writing reports with litigation in mind.

5 Gowers, Sir Ernest, *The complete plain words*, revised by Sidney Greenbaum and Janet Whitcut. London, HMSO, 3rd edition 1986. Indispensable if clarity and simplicity are objectives.

Study 4
Remedial work

1 Introduction

1.01 Scope and nature variable
It is difficult to provide detailed or comprehensive guidance on either the scope or nature of remedial work for specific cases (but see 'Note' later). So much depends on: the precise nature of the defect and its context; the life expectancy required of the remedy; the cost of disturbance to or rehousing of occupiers during the course of the remedial work; and, always important, the amount of money available for the remedial work. However, there are some useful general principles that might be applied to specific cases. These are listed below.

2 General principles

2.01 Accurate diagnosis
To be successful – and this may be all too obvious – remedial work must be based on as accurate a diagnosis of the cause of the defect as is reasonably possible. Unless this is done, the remedial work may not only turn out to be quite inappropriate or short-lived but may also lead to other defects.

2.02 Acceptable degree of alteration
The extent to which the existing appearance of building can or should be altered by the remedial work should be considered at a very early stage, **1**.

2.03 Clear understanding of underlying principles
Accepted good practice, including the latest advice resulting from reliable research work, should provide the basic guidance, but often some compromises because of practical difficulties or cost constraints are inevitable. Care is therefore needed when devising appropriate solutions to ensure that the principles underlining good practice are rigorously applied.

2.04 Hazards of multiple-layer additions
Where it is necessary to add additional layers of materials to an existing construction, special care is required to ensure that the type of material and its location especially do not induce a risk of interstitial condensation, other deleterious thermal effects or entrapment of moisture.

2.05 Stage-by-stage approach
In those cases where interacting factors are complex (as in condensation problems, for example) it is often more practical (and sometimes cheaper in the long run) to carry out remedial work in stages, having first established a comprehensive remedial plan. This stage-by-stage approach should be carried out on an experimental basis.

Note
In Part 2 more specific guidance on remedial work is given for each defect included in the studies. More comprehensive guidance on a wider range of defects than covered in this book is given in: Eldridge, H. J., *Common defects in buildings*, London, HMSO, 1976. Systematic and authoritative. Well illustrated. Specific guidance on remedial work in each case. The cases covered by Eldridge include the more 'traditional' kind. Eldridge's book has been replaced by *Defects in buildings*, DOE, PSA Directorate of Building Development, London, HMSO, 1989. The new work includes but updates Eldridge's cases using a slightly different framework and adds many new cases and problems.

1 *Remedial work may cause unacceptable changes to the appearance of a building*

Part 2: Failures in context

Study 5
Introduction

1 The form of presentation

This part provides the basis for the identification of a defect, its diagnosis and remedy; with the emphasis placed on diagnosis. The commonly occurring failures are grouped according to elements and components and the information related to them is presented under a recurring set of headings as follows.

1.01 Key diagram
This gives the relevant part of the building, **1**.

1.02 Manifestation
The way or ways in which the defect has manifested itself is/are described and, wherever possible, illustrated with appropriate photographs.

1.03 Condition
The drawings included here, which are mostly three-dimensional, are intended to illustrate the constructional details that have led to the defect arising. The conditions chosen for inclusion represent generalised details for the defect under consideration, but it should be possible to relate these, without too much error, to other basically similar conditions that have given rise to similar defects, provided the principles involved are interpreted appropriately.

1.04 Cause
The most likely cause(s) and their sources for the defect are included here in expanded note form. In most cases alternatives are given. In general, the notes given are intended to provide *basic* guidance on the factors that require detailed investigation. Further general guidance on this aspect is given in Study 3, 'Diagnosis: principles and procedures', and in Part 3 of this book.

Note on causes
Account *must* be taken of the fact that only those causes of relevance to the defects described in these studies are included. Wherever possible, a brief note on other possible causes of similar symptoms is added.

1.05 Cross references
Reference is given here mainly, but not exclusively, to those sections of Part 3 that contain 'diagnostic checklists' which should aid the detailed investigation into the causes and sources of the defect.

1.06 Remedial work
General guidance is given again in expanded note form on the nature and scope of the remedial work most likely to be required. Alternative courses of action are noted. The advice given here should be read in conjunction with Study 4, 'Remedial work' in Part 1. It is important to stress the

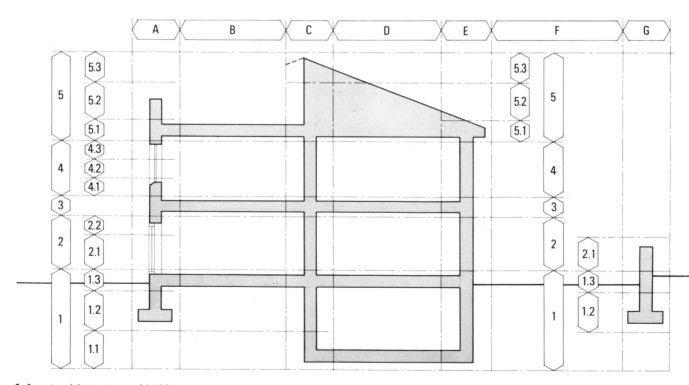

1 *Locational key to parts of building.*

need for each problem to be solved on its merits. When devising remedial measures it will be necessary to seek out (and the search is likely to be repaid) special fittings and fastenings that are made specifically with remedial work in mind.

1.07 References for correct detailing

Those external references included here (and they have been carefully selected) are intended primarily to guide architects when undertaking the detailing of new work; but they will also be useful in providing guidance on the details that should be archieved in the remedial detail. In some cases reference is given to details specifically devised by others to remedy certain kinds of defects. A guide to the content, scope and usefulness of most of the references will be found in the annotated reading lists in the appropriate sections of the studies in Part 3. For convenience, the references included in Studies 6–19 that follow are summarised at the end of this study.

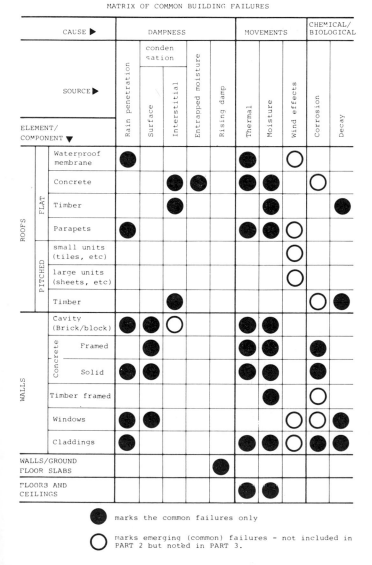

2 *Matrix showing interrelationship between building parts in which failure most commonly occurs, and causes/sources of failure. Studies in Part 2 are structured along the vertical axis of matrix; studies in Part 3 along the horizontal axis.*

2 Relationship between Parts 2 and 3

2.01

The relationship between Studies 5–19 which constitute Part 2, and Studies 20–27 which constitute Part 3, is shown on the matrix, **2**. The physical elements of the building in which failure will manifest itself are arranged vertically at the left (this is the architect's point of entry) and provides the structure for the set of information sheets. The range of possible causes and sources of defects is displayed horizontally, and this provides the structure for the concluding set of technical studies.

The junctions representing the most commonly occurring failures are marked with solid circles; less common ones (noted in the studies in Part 3 but not in the studies in Part 2) are marked with an open circle.

Deducting the likely cause and source of a defect will therefore involve careful checking back and forth between the relevant study in Part 2 (say, flat roof: timber) and the various studies in Part 3 which may provide the explanation of failure (say, condensation; movement; and decay).

3 Summary of references

3.01

In the studies included in Part 2, appropriate external sources for detailed information are referred to systematically. For convenience, the recurring references are summarised below: readers/users of the studies should acquire, or have access to, the full set.

3.02 British Standards and Codes of Practice

These are available from BSI Sales Department, Linford Wood, Milton Keynes, MK14 6LE (0908 221166). Becoming a subscribing member ensures up-to-date information via the monthly *BSI News* and reduces rates. Before using a Standard or Code of Practice make sure that the latest amendments are included. The list below includes amendments given in the BSI Standards Catalogue, an annual publication.

Note: Codes of Practice (CPs), mostly in A5 size, are being phased out. New or revised Codes of Practice are now included as BSs.

Standards
BS 0743: 1970
Specification for materials for damp-proof courses including AMD 2503 (March 1987), AMD 4336 (September 1983) and AMD 4594 (October 1984). 1970

BS 0747: 1977 (1986)
Specification for roofing felts including AMD 3775 (November 1981), AMD 4606 (February 1985) and AMD 5101 (February 1986). 1986

BS 0988,1076,1097,1451: 1973
Specification for mastic asphalt for building (limestone aggregate) (replaces BS 988,1097,1076,1451: 1966)

BS 2871: Part 1: 1971
Specification for copper and copper alloys. Part 1: Copper tubes for water, gas and sanitation including AMD 1422 (May 1974) and AMD 2203 (November 1976). 1971

BS 3712: Parts 1, 3, 4: 1985; Part 2: 1986
Building and constructional sealants

BS 3826: 1969
Specification for silicone-based water-repellents for masonry including AMD 2141. October 1976.

BS 3921: 1985
Specification for clay bricks. November 1985

BS 4254: 1983
Specification for two-part polysulphide-based sealants, including AMD 5023 (December 1985). 1983

BS 4255: Part 1: 1986
Rubber used in preformed gaskets for weather exclusion from buildings: Part 1 Specification for non-cellular gaskets (Part 2: 1975 Cellular gaskets withdrawn). 1986

BS 5215: 1986
Specification for one-part gun grade polysulphide-based sealants including AMD 5468 (September 1986). 1986

BS 5247: Part 14: 1975
Code of practice for sheet roof coverings, corrugated asbestos cement. 1975

BS 5250: 1989
Code of practice for control of condensation in buildings. (A major revision of BS5250: 1975 together with a change of title – buildings rather than dwellings only now covered).

BS 5262: 1976
Code of practice for external rendered finishes (replaces CP 221:1960) including AMD 2103 (September 1976). September 1976

BS 5350
Methods of test for adhesives in various parts – some still not published

BS 5385: Part 1: 1976
Code of practice for internal ceramic wall tiling and mosaics in normal conditions including AMD 5180 (June 1986). 1976

BS 5383: Part 2: 1978
Code of practice for external ceramic wall tiling and mosaics (replaces CP 212: Part 1: 1963) including AMD 5181 (June 1986). 1978

BS 5385: Part 4: 1986
Code of practice for ceramic tiling and mosaics in specific conditions.

BS 5442: 1977–1979
Classification of adhesives for construction. In three parts. 1977–1979

BS5427: 1976
Code of practice for performance and loading criteria for profiled sheeting in building.

BS 5534: Part 1: 1978 (1985)
Code of practice for slating and tiling Part 1. Design preplaces CP 142: Part 2). Latest amendment is in line with latest wind code (April 1985)

BS 5589: 1978
Code of practice for preservation of timber including AMD 3916 (April 1982). 30 September 1978

BS 5606: 1990
Guide to accuracy in building. (Change from a 'Code' to a 'Guide').

BS 5618: 1978
Code of practice for thermal insulation of cavity walls (with masonry inner and outer leaves) by filling with ureaformaldehyde (UF) foam. 31 August 1978

BS 5628: Part 3: 1985
Code of practice for use of masonry Part 3. Materials and components, design and workmanship (formerly CP 121: Part 1) including AMD 4974 (November 1985). 29 March 1985

BS 5642: Part 1: 1978
Sills and copings: Part 1. Specification for window sills of precast concrete, cast stone, clayware, slate and natural stone (replaces BS 4374). 30 October 1981

BS 5642: Part 2: 1983
Sills and copings: Part 2. Specification for copings of pre-cast concrete, cast stone, clayware, slate and natural stone. (Replaces BS 3798: 1964.) 1983

BS 6093: 1981
Code of practice for design of joints and jointing in building construction. 1981

BS 6150: 1982
Code of practice for painting of buildings (replaces CP 231: 1966). 31 May 1982

BS 6213: 1982
Guide to selection of constructional sealants including AMD 5466 (April 1987). 31 May 1982

BS 6229: 1982
Code of practice for flat roofs with continuously supported coverings. 31 December 1982

BS 6262: 1982
Code of practice for glazing and fixing of glass for building (replaces CP 152) including AMD 4063 (June 1982) and AMD 4582 (June 1984). 30 June 1982

BS 6375: Part 1: 1983
Performance of windows: Part 1. Classification for weathertightness (including guidance on selection and specification) (replaces DD 4: 1971) including AMD 4907 (July 1985). 29 April 1983

BS 6477: 1984
Specification for water repellents for masonry surfaces. 1984

BS 6576: 1985
Code of practice for installation of chemical damp-proof courses. 28 February 1985

BS 6577: 1985
Specification for mastic asphalt for building (natural rock asphalt aggregate). (Replaces BS 1162,1418,1410: 1973.)

BS 6676: Part 2: 1986
Code of practice for installation of batts (slabs) filling the cavity construction. 1986

BS 6682: 1986
Method for determination of bimetallic corrosion in outdoor exposure corrosion tests. 1986

BS 8102: 1990
Code of practice for protection of structures against water from the ground. 1990. Partially replaces CP 102: 1973.

BS 8110: Part 1: 1985
Structural use of concrete: code of practice for design and construction. 1985

BS 8200: 1985
Code of practice for design of non-loadbearing external vertical enclosures of buildings. 28 June 1985

BS 8208: Part 1: 1985
Guide to assessment of suitability of external cavity walls for filling with thermal insulants. Part 1. Existing traditional cavity construction. 31 July 1985. AMD 4996, September 1985

Codes of practice
CP 102: 1973
Code of practice for protection of buildings against water from the ground including AMD 1511 (July 1974), AMD 2196 (January 1977) and AMD 2470 (February 1978). 1973

Partially replaced by BS 8102: 1990

CP 143: Parts 1–16, 1958–1974
Code of practice for Sheet roof and wall coverings.

CP 144: Part 3: 1970
Roof coverings Part 3: Built-up bitumen felt. Metric units including AMD 2527 (April 1978) and AMD 5229 (June 1986) (being revised). 1970

CP 144: Part 4: 1970
Roof coverings Part 4: Mastic asphalt. Metric units. 1970

Drafts for development
DD 93: 1984
Draft for Development: Methods of assessing exposure to wind-driven rain. 30 March 1984

Published document
PD 6484: 1984
Commentary on corrosion at bimetallic contacts and its alleviation. 1984

3.03 BRE publications
Almost all of BRE's publications are available from Publications Sales Office, Building Research Establishment, Garston, Watford, Herts WD2 7JR (0923 674040). The best way to keep up to date on Digests, Defect Action Sheets, Information Papers and the annual Information Directory is to take out an annual subscription. For the complete package this is £70 (1991 price).

Digests
8   Built-up felt roofs. New edition, 1970
33 Sheet and tile flooring made from thermoplastic binders. Revised, 1971
35 Shrinkage of natural aggregates in concrete. New edition 1968, revised 1971
54 Damp-proofing solid floors. New edition, 1971
69 Durability and application of plastics. New edition 1973, revised 1977
75 Cracking in buildings. Reprinted 1975. Superseded by Digest 361 but retained here as it still contains a worthwhile classification.
77 Damp-proof courses. Revised, 1971
79 Clay tile flooring. Revised, 1976
95 Durability of metals in natural waters. Revised, 1977
104 Floor screeds. New edition 1973, revised 1979
109 Zinc-coated reinforcement for concrete. September 1969
110 Condensation. Revised, 1972
127 An index of exposure to driving rain. Revised, 1971
130 (First series) Asbestos sheets cracking
137 Principles of joint design. New edition, 1977
139 Control of lichens, moulds and similar growths. New edition, 1977

144 Asphalt and built-up felt roofing: durability. August 1972
152 Repair and renovation of flood-damaged buildings. 1973
157 Calcium silicate (sandlime, flintlime) brickwork. New edition, 1981
161 Reinforced plastics cladding panels. January 1974
163 Drying out of buildings. March 1974
196 External rendered finishes. December 1976
197 Painting walls: Part 1: Choice of paint. New edition, 1982
198 Painting walls: Part 2: Failures and remedies. Revised, 1984
200 Repairing brickwork (April 1977). Minor revisions, 1981
201 Wood preservatives: application methods. May 1977
211 Use of site adhesives: Part 1. March 1978
212 Use of site adhesives: Part 2. April 1978
213 Choosing specifications for plastering. May 1978
217 Wall cladding defects and their diagnosis. 1978
218 Cavity barriers and ventilation in flat and low pitched roofs. 1978
223 Wall cladding: designing to miminise defects due to inaccuracies and movements. March 1979
227 Estimation of thermal and moisture movements and stresses: Part 1. July 1979
228 Estimation of thermal and moisture movements and stresses: Part 2. August 1979
229 Estimation of thermal and moisture movements and stresses: Part 3. September 1979
231 Specifying timber, August 1984
236 Cavity insulation (April 1980). Minor revisions, 1984
245 Rising damp in walls: diagnosis and treatment. Revised, 1981
250 Concrete in sulphate-bearing soils and groundwaters. Revised, 1986
263 The durability of steel in concrete: Part 1 Mechanism of protection and corrosion. July 1982
264 The durability of steel in concrete: Part 3 The repair of reinforced concrete. August 1982
265 The durability of steel in concrete: Part 2 Diagnosis and assessment of corrosion-cracked concrete. September 1982
270 Condensation in insulated domestic roofs. February 1983
273 Perforated clay bricks. May 1983
277 Built-in cavity insulation for housing. September 1983
296 Timbers: their natural durability and resistance to preservative treatment. April 1985
299 Dry rot: its recognition and control. July 1985
301 Corrosion of metals by wood. September 1985
304 Preventing decay in external joinery. December 1985
305 Zinc-coated steel. January 1986
312 Flat roof design: the technical options (replaces Digest 221). August 1986
321 Timber for joinery. May 1987
325 Concrete Part 1: materials (replaces Digests 237 and 244). October 1987
326 Concrete Part 2: specification, design and quality control (replaces Digests 237 and 244). October 1987
329 Installing wall ties in existing construction, Feb. 1988
336 Swimming pool roofs: minimising the risk of condensation using warm-deck roofing, September 1988
340 Choosing wood adhesives, January 1989
343 Simple measuring and monitoring of movement in low-rise buildings, Part 1: cracks, April 1989

344 Simple measuring and monitoring of movement in low-rise buildings, Part 2: settlement, heave and out-of-plumb, May 1989.

345 Wet rots: recognition and control, June 1989

354 Painting exterior wood (replaces Digest 261), September 1990

361 Why do buildings crack (replaces Digest 75)

362 Building mortar (Replaces Digests 89 and 160)

Defect Action Sheets (DAS)

1 Slated or tiled pitched roofs: ventilation to outside air. May 1982

3 Slated or tiled pitched roofs: restricting the entry of water vapour from the house. June 1982

4 Pitched roofs: thermal insulation near the eaves. June 1982

9 Pitched roofs: sarking felt underlay – drainage from roof. November 1982

10 Pitched roofs: sarking felt underlay – water-tightness. November 1982

16 Walls and ceilings: remedying recurrent mould growth. January 1983

17 External masonry walls insulated with mineral fibre cavity-width batts: resisting rain penetration. February 1983

18 External masonry walls: vertical joints for thermal and moisture movements. February 1983, first edition; new edition, February 1985

19 External masonry cavity walls: wall ties – selection and specification. February 1983

20 External masonry cavity walls: wall ties – installation. February 1983

21 External masonry cavity walls: wall ties – replacement. February 1983

22 Ground floors: replacing suspended timber with solid concrete – dpcs and dpms. March 1983

35 Substructure: dcps and dpms – specification. New edition, March 1985

36 Substructure: dcps and dpms – installation. September 1983

37 External walls: rendering – resisting rain penetration. October 1983

38 External walls: rendering – application. October 1983

59 Felted cold deck flat roofs: remedying condensation by converting to warm deck. September 1984

67 Inward-opening external doors: resistance to rain penetration. 1985

68 External walls: joints with windows and doors – detailing for sealants. December 1985

69 External walls: joints with windows and doors – application of sealants. December 1985

70 External masonry walls: eroding mortar – repoint or rebuild? February 1986

71 External masonry walls: repointing – specification. February 1986

72 External masonry walls: repointing. February 1986

75 External walls: brick cladding to timber frame – the need to design for differential movement. April 1986

76 External walls: brick cladding to timber frame – how to allow for movement. April 1986

77 Cavity external walls: cold bridges around windows and doors. May 1986

79 External masonry walls: partial cavity fill insulation – resisting rain penetration. June 1986

85 Brick walls: injected dpcs. August 1986

86 Brick walls: replastering following dpc injection. August 1986

94 Masonry chimneys: dpcs and flashings – location. February 1987

95 Masonry chimneys: dpcs and flashings – installation. February 1987

97 Large concrete panel external walls: resealing butt joints. March 1987

106 Cavity parapets – avoiding rain penetration. August 1987

107 Cavity parapets – installation of copings, dpcs, trays and flashings. August 1987

115 External cavity walls: wall ties – selection and specification, June 1988

116 External cavity walls: wall ties – installation, June 1988

133 Solid external walls: internal dry-lining – preventing summer condensation, June 1989.

Note: Defect Action Sheets have been discontinued and a new series, Good Building Guides, started in April 1990.

Good Building Guides (GBG)
This series, started in April 1990, aims to provide practitioners with concise guidance on the principles and practicalities of achieving good quality building. Those currently available are extremely well illustrated in colour and mostly in three-dimensions.

3 *Damp proofing basements*, July 1990

5 *Choosing between cavity, internal and external wall insulation*, October 1990

Current papers

CP 36/64, *Demands on rubbers in buildings*, by G. W. Mack. 1964

CP 86/74, *Window wall joints*, by M. R. M. Herbert. 1974

CP 94/74, *The rippling of thin flooring over discontinuities in screeds*, by J. Warton and P. W. Pye. 1974

CP 7/75, *The independent core method – a new technique for the determination of moisture content*, by A. J. Newman. 1975

CP 3/81, *The performance of cavity wall ties*, by J. F. A. Moore. 1975

Information papers

IP 28/79, *Corrosion of steel wall ties: recognition, assessment and appropriate action*, by R. C. de Vekey. October 1979

IP 35/79, *Moisture in a timber-based flat roof of cold deck construction*, by I. S. McIntyre. November 1979

IP 10/80, *Avoiding joinery decay by design*, by Janice K. Carey. July 1980

IP 4/81, *The performance of cavity wall ties*, by J. F. A. Moore. April 1981

IP 7/83, *Window to wall jointing*, by Marilyn J. Edwards. May 1983

IP 15/83, *Emulsion-based formulations for remedial treatments against woodworm*, by R. W. Berry and R. J. Orsler Morgan. October 1983

IP 4/84, *Performance specifications for wall ties*, by R. C. de Vekey. March 1984

IP 6/84, *The movement of foam plastics insulants in warm deck flat roofs*, by J. C. Beech and G. K. Saunders. April 1984

IP 8/84, *Ageing of wood adhesives – loss in strength with time*, by D. F. G. Rodwell. July 1984

IP 11/85, *Mould and its control*, by A. F. Bravery. June 1985

IP 3/86, *Changes in Portland cement properties and their effects on concrete*, by P. J. Nixon. March 1986

IP 13/87, *Ventilating cold deck flat roofs*, by J. C. Beech and S. Uberoi. October 1987

IP 20/87, *External joinery: end grain sealers and moisture control* by E. R. Miller, J. Boxall and J. K. Carey, December 1987

IP 2/88, *Rain penetration of cavity walls: report of a survey of properties in England and Wales* by M. T. Pountney, BA, R. Maxwell, BSc, MSc and A. J. Butler, MPhil, CEng, MIMechE, February 1988

IP 12/88, *Summer condensation on vapour checks: tests with battened, internally insulated, solid walls* by J. R. Southern, November 1988

IP 16/88, *Ties for cavity walls: new developments* by R. C. deVekey, PhD, CChem, DIC, MRSC

IP 17/88, *Ties for masonry cladding* by R. C. deVekey, PhD, CChem, DIC, MRSC, December 1988

IP 2/89, *Thermal performance of lightweight inverted warm deck flat roofs* by J. C. Beech and G. K. Saunders, January 1989

IP 7/89, Davies H. and Rothwell, G. W., *The effectiveness of surface coatings in reducing carbonation in reinforced concrete*, May 1989

BRAS Technical Information Leaflets (TIL)

TIL 5: 1971, *Devices for detecting changes in width of cracks in buildings*. (Covers simple mechanical devices. More sophisticated versions now available commercially.) 1970

TIL 14: 1971, *Shelling of plaster finishing coats*. 1971

TIL 19: 1974, *Pitting corrosion of copper tubing*. 1974

TIL 22: 1974, *Corrosion of wall ties in cavity brickwork*. 1974

TIL 24: 1977, *Plastering on dense concrete*. 1977

TIL 29: 1977, *Diagnosing rising damp*. 1977

TIL 36: 1972, *Chemical damp-proofing courses for walls*. 1972

TIL 39: 1971, *Thin walled steel tubing in water services*. 1971

TIL 55: 1977, *Cracking of rendered lightweight concrete block walls*. May 1977

TIL 59: 1977, *Condensation in domestic tiled pitched roofs – advice to householders*. 1977

TIL 65: 1981, *Window materials and finishes*. 1981

TIL 66: 1981, *Treatment of mould – advice to householders*. October 1981

Reports

BRE Report: *Directional driving rain indices for the United Kingdom – computation and mapping* (background to BSI Draft for Development DD93), by M. J. Prior. DOE, BRE, 1985

BRE Report: *Overcladding external walls of large panel system dwellings*, by H. W. Harrison, J. H. Hunt and J. Thomson. DOE, BRE, 1986

BRE Report: *Driving-rain index*, by R. E. Lacy. DOE, London, HMSO, 1976

BRE Report: *Rain penetration through masonry walls: diagnosis and remedial measures* by A. J. Newman, BRE, 1988

BRE Report: *Thermal insulation: avoiding risks*, BRE/HMSO, 1989

3.04 Books, trade and government publications

Books

Addleson, Lyall and Rice, Colin, *Performance of materials in buildings*, Butterworth-Heinemann, 1992. A substantial revision and overhauling of Lyall Addleson's *Materials for building*, Vols 1–3 (excluding 1 Physical and chemical aspects of matter, 2.01 Elementary considerations of strength and 2.02 Mechanical properties in Vol 1) with a full explanation of 'Principles for building'.

Addleson, Lyall, *Materials for building*, Vol 4, Newnes/Butterworths, 1978

Anderson, J. M. and Gill, J. R., *Rainscreen cladding*, a guide to design principles and practice, CIRIA/Butterworths, 1988.

Brookes, A. J., *Cladding of buildings*, London, Construction Press, 1983

Brookes, Alan J., *Concepts in cladding: case studies on jointing for architects and engineers*, London, Construction Press, 1985

Brookes, A. J. and Grech, C. *The Building Envelope, applications of new technology cladding*, Butterworth Architecture, 1990

Butler, G. and Ison, H. C. K., *Corrosion and its prevention in waters*, London, Leonard Hill, 1966

Cartwright, K. St G. and Findlay, W. P. K., *Decay of timber and its prevention*, 2nd edition, London, HMSO, 1958

DOE, *Condensation in dwellings. Part 2: Remedial work*, London, HMSO, 1971

Desch, H. E., revised by J. M. Dinwoodie, Princes Risborough Laboratory, *Timber, its structure and properties*, 6th edition, London, Macmillan, 1981

Duell, John and Lawson, Fred, *Damp proof course detailing*, 2nd edition, London, Butterworth Architecture, 1983

Eldridge, H. H., *Properties of building materials*, Lancaster, MTP Construction, 1974

Gibson, E. J. (ed.), *Developments in building maintenance – 1*, London, Applied Science, 1979

Handisyde, Cecil, *Everyday details*, London, Butterworth Architecture, 1979

March, Francis, *Flat roofing. A guide to good practice*, Tarmac Building Products Limited, 1982

Marsh, P., *Fixings, fasteners and adhesives*, London and New York, Construction Press, 1984

Martin, Bruce L., *Joints in buildings*, London, George Godwin, 1977

Nash, W. G., *Brickwork repair and restoration*, Eastbourne, Attic Books, 1986

PSA *Technical guide to flat roofing*, 1987, DOE/PSA, March 1987. By its new title, it replaces *PSA Flat roofs, Technical guide, 2nd edition*, PSA, December 1981. (It is loose-leaf in two volumes, diagrams in colour and format much altered. Biased towards the inverted roof with new rules to those in *Inverted roofs – technical guide*, see below). *Note:* The second edition should be retained as a reference for its treatment of maintenance and for its annexes on testing.

Richardson, Barry A., *Remedial treatments of buildings*, Lancaster, The Construction Press, 1980

Journals

AJ Technical
Controlling the risk, Code for condensation, by Peter Burberry, 26.09.90, pp. 59–63

AJ Energy File
Condensation and how to avoid it, by Peter Burberry, 3.10.79, pp. 723–739

AJ Series
Element design guide: External walls 1 Masonry: Part 2, 9.07.86. 3 Curtain walls, 23.07.86. 4 Profiled metal sheets, 30.07.86. 5 Metal panels, 6.08.86. 7 Precast concrete, 13.08.86

AJ Series
Element design guide: Roofs 3 Slate and tile roofs, 25.03.87. 4 Profiled sheet roofs, 1.04.87. 7 Flat roof construction, 22.04.87

AJ Series
Construction risks and remedies: Condensation, 9 and 16.04.86. Thermal insulation, 11 and 18.06.86, Timber decay, 8 and 15.10.86

AJ Series
Products in practice – now superseded by *AJ Focus*

Building Technical File
BRAS, Cracking and bulging of plaster, 4, pp. 45 and 46, January 1984

Building Design: Brick Supplement
Water off a duck's back (or how not to sink like a brick), by Donald Foster (BDA), 1983

Building Technical File
PSA Feedback, Single storey extensions – cavity wall defects, 2, 43 (July 1983)

Building Technical File
Fibreglass Ltd, Cavity wall test facilities, by Rick Wilberforce, 3, pp. 51–53 (October 1983)

Building Technical File
BRE, Failures with site-applied adhesives, by J. R. Coad and D. Rosaman, 7, pp. 57–63 (October 1983)

Building Technical file
Pilkington Research and Development, The calculation of interstitial condensation risk, by K. A. Johnson, 26, pp. 23–30 (July 1989)

Building Technology and Management
Selection, performance and replacement of building joint sealants, by J. C. Beech (BRE), Part 1, 21(7), pp. 23–25 and Part 2, 21(8), pp. 14–16

Construction
Formerly *DOE Construction,* published by DOE. Now published by B&M Publications (London) Limited, PO Box 13, Hereford House Bridle Path, Croydon CXR9 4NI (01-680 4200)

DOE, *Construction*
Quarterly. Formerly published by the DOE. Now published by B&M Publications (London) Limited as *Construction*

GLC, *Development and Materials Bulletins*
Formerly published by the now extinct GLC. Some items in *Building Technical File* until the GLC's dissolution. (Source of back numbers unknown)

The Architect
Light cladding systems, by Barry Josey, pp. 61–81 (September 1987)

Trade
ACBA, BDA, C&CA, Eurosil-UK, NCIA and STA
Cavity insulated walls, Practice Note 1 (September 1984)

ACBA, BPF, BDA, C&CA, Eurosil-UK and NCIA
Cavity insulated walls, Specifiers guide (January 1987)

British Flat Roofing Council (formerly Bituminous Roofing Council (BRC)
Publishes technical information on flat roofs. Felt roof and information centre, BFRC, 38 Bridlesmith Gate, Nottingham, NG1 2GQ Tel: (0602) 507733/Fax: (0602) 504122

Brick Development Association (BDA)
Publishes a range of technical information including Technical Notes referred to in studies. BDA, Woodside House, Winkfield, Windsor, Berks SL4 2DX (0344 885651)

British Steel Strip Products
Profiled sheet steel roofs and internal humidity, April 1988. Enquiries to Steel Strip Products, PO Box 10, Newport Gwent, NP9 0XN (Tel: 0633 290022/Fax: 0633 272933)

C&CA
Appearance matters 2
External rendering, by William Monks and Fred Ward (1982) Enquiries to British Cement Association (BCA), Wexham Springs, Slough, Berks, SL3 6PL (028 162727)

Copper Development Association
Orchard House, Mutton Lane, Potters Bar, Herts EN6 3AP (0707 50711)

Felt Roofing Contractors' Advisory Board (FRCAB)
Roofing handbook, June 1988 (Loose leaf sheets, replacing *Built-up roofing,* revised). Enquiries to Maxwelton House, Boltro Road, Haywards Heath, West Sussex RH16 1BJ (0444-51835) (June 1982)

Lead Development Association
34 Berkeley Square, London W1X 6AJ (071-499 8422)

Mastic Asphalt Council and Employers Federation
MACEF Roofing handbook (revised edition) (April 1985). Technical enquiries to MATAC, Lesley House, 6–8 The Broadway, Bexleyheath, Kent DA6 7LE Tel: 081-298 0414/Fax: 081-298 0387

NHBC
Practice Note 13, *Construction of flat and pitched roofs with fully supported continuous weatherproof membrane,* NBHC, Chiltern Avenue, Amersham, Bucks HP6 5AP

National Federation of Roofing Contractors (NRFC)
Profiled sheet metal roofing cladding (May 1982)

Sealant Manufacturers' Conference/CIRIA
Manual of good practice in sealant application. Enquiries to Sealant Manufacturers' Conference, 15 Tooks Court, London EC4 1LA (071-831 7851) or CIRIA (January 1976)

Timber Research and Development Association (TRADA)
Stocking Lane, Hughenden Valley, High Wycombe, Bucks HP14 4ND (0240 243091/Fax: 0240 245487)

Zinc Development Association
34 Berkeley Square, London W1X 6AJ (071-499 6636)

British Board of Agrément
Agrément Certificates for independent technical appraisal and information on new building materials and products. Enquiries to BBA, PO Box 195, Bucknalls Lane, Garston, Watford, Herts WD2 7NG (0923 662900)
BBA and others, *Cavity insulation of masonry walls – dampness risks and how to miminise them* (November 1983)
Information sheet, *Cavity wall insulation*
Information sheet 10, *Methods of assessing the exposure of buildings for cavity fill insulation* (revised edition) (January 1983)
Information sheet 16, *Cavity wall insulation* (January 1982)

Study 6
Flat roofs: waterproof membranes

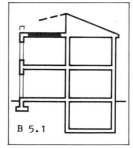

B 5.1

Manifestation 1, Conditions 1 and 2

1 Crazing of the asphalt surface that is not causing ingress of water.
2 Wrinkling of asphalt surface that is not causing ingress of water

Cause, Condition 1

1 The sand used to mop up bitumen brought to the surface during laying was not rubbed in while the asphalt was still warm from laying; or
2 Solar reflective treatment is absent or has failed, **6**; or
3 There is or has been ponding of rainwater or melting snow on the roof, **5**. The ponding may have been due to inadequate falls when the roof was laid; slight deflections below the general level of the drainage, or the rainwater outlets were set too high.

Remedial work

Technically, no remedial work is necessary. Appearance would be the only criterion for treating the asphalt. However, the asphalt life will be prolonged if a solar reflective treatment is applied where there is none.

If the cause of the ponding is due to incorrectly formed rainwater outlets (i.e. the outlets are too high or the asphalt around them has been laid too high) then the outlets should be reformed. Rainwater outlets cast into concrete roof slabs will present difficulties; each case should be considered in the light of particular circumstances.

Cause, Condition 2

Restraint of the thermal movement of asphalt due to either:

1 Inadequate separation of asphalt from its base; or

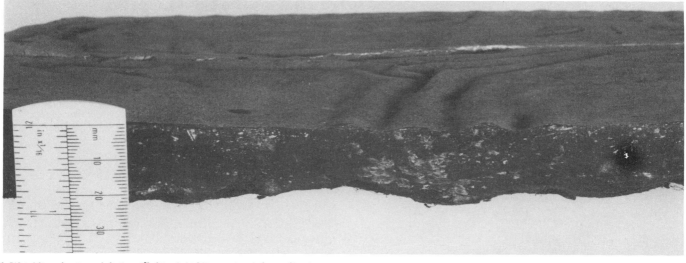

1 *Wrinkling due to undulating (lightweight bitumen bonded screed) substrate for the asphalt. Note uneven thickness of asphalt and the depth of the wrinkles.*

2 *and* **3** *Wrinkling of asphalt around perimeter of laid* in situ *screed paving laid on polythene separating layer. The latter is wrinkled by the wet screed, the wrinkles subsequently being imprinted on the underside of the asphalt thereby causing the latter to be restrained from moving.* **2** *general view;* **3** *at corner with polythene sheet lifted up.*

2 Asphalt laid on a rough or undulation surface, **1**; or
3 Inadequate separation of laid *in-situ* cement-based paving laid on the surface of the asphalt, **2**, **3**;
4 Solar reflective paints containing aluminium flakes, **4**.

Remedial work

Technically, no remedial work is necessary provided the wrinkles are not deep enough to induce nor to have caused cracking of the asphalt.

If the wrinkling is local only and due to causes 1 to 3 above, the paving/asphalt should be reinstated locally around the cause of the restraint, otherwise the asphalt should be reinstated completely.

If the cause is the type of solar reflective paint used, the wrinkling may be arrested and the life of the asphalt prolonged if the paint is removed (without damaging the asphalt) and chippings used as the solar reflective treatment.

References for correct detailing
1 For sanding warm asphalt:
  (a) CP 144: Part 4: 1970: *Roof coverings: mastic asphalt*, clause 4.6.2, page 24, requires that immediately after the required number of coats has been laid, clean sharp sand should be rubbed into the surface of the asphalt using a wooden float.
  (b) Tarmac's *Flat roofing, a guide to good practice*, 4.2 Asphalting techniques (pages 118 and 119, the end of the latter describing the sand rubbing).
2 For solar reflective treatments;
  (a) CP 144: Part 4: 1970: *Roof coverings: mastic asphalt*, clause 3.11 suggests:
   ● Light-coloured chippings of limestone, granite, gravel, calcite or felspar 6–10 mm in size, set in bitumen dressing compound immediately after laying the asphalt.
   ● Concrete or asbestos cement tiles bedded in bitumen bonding compound.

4 *Wrinkling associated with solar reflective paint containing aluminium flakes.*

6 *Surface wrinkling and crazing produced by solar radiation. Again, cracks are small and shallow and difficult to capture photographically. (Photo GLC.)*

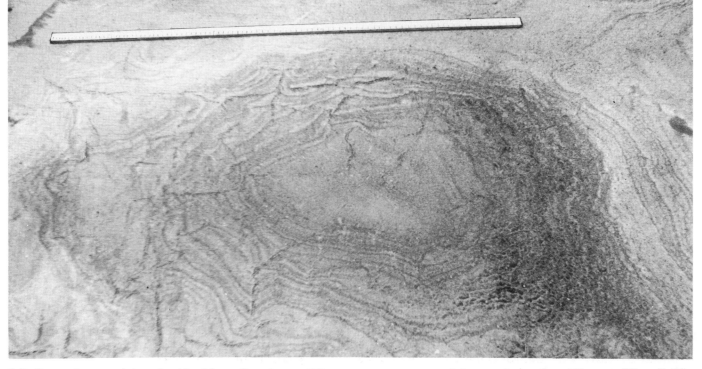

5 *Surface crazing on asphalt roof, produced by ponding of water. Effect is difficult to capture photographically, as crazing involves small surface cracks,* *not penetrating asphalt more than a few millimetres. (Photo Building Research Establishment, Crown Copyright)*

● Cement/sand screed, 25 mm thick, cut into small squares (not more than 600 mm) with a 75 mm margin at the edges for movement. The edge margin should be filled with a bitumen compound.

Note: The CP advises against the use of solar aluminium-based reflective paints as permanent protection but see 2(c) below.

(b) PSA, *Technical guide to flat roofing*, 1987, 3.12 Protective layer, pp. 57–58 tabulates all types of protective layers, including those in CP 144 (a) above, but is biased towards the inverted system.

(c) GLC, *Development and Materials Bulletin*, **119**, October 1978, item 1, 'A durability trial of solar reflective paints on asphalt' and **120**, December 1979, item 3, 'Solar reflective coatings on asphalt – their effectiveness.' Solar reflective paints will give a temperature reduction of between 15 and 35 per cent, depending on the age of the coat and the amount of dust on it. The paint coating will not prevent blisters or slumping but will help to reduce the severity of these defects. Light-coloured chippings are the most effective treatment and where these are used reflective paints should be applied to the skirtings where the chippings cannot be laid.

Note: MACEF advises its members against the use of solar reflective paints due mainly to the need for continual maintenance of the coating.

Some aluminium-based solar reflective paints appear to cause much wrinkling of the surface of asphalt. Why this should be so has yet to be explained conclusively.

(d) Tarmac's *Flat roofing, a guide to good practice*, 4.3. Surface protection, pages 120 and 121, discusses the merits of different ways of achieving surface protection of asphalt.

3 For avoiding ponding:

(a) PSA, *Technical guide to flat roofing*, 1987, 2.9 Falls and drainage, pp. 27–29, explains why falls should be designed to 1 in 40 for achievement in practice of recommended falls of 1 in 80 and relationship between falls and outlets.

(b) CP 144: Part 4: 1970: *Roof coverings: mastic asphalt*, Figs 16 and 17 on pages 42 and 43 illustrate clearly the asphalt/rainwater outlet junction for lead and cast iron. The dishing of asphalt around the outlets should be noted and achieved. The code gives no recommendations for plastics rainwater outlets – if these are to be used, consult the manufacturer for his recommendations.

(c) Tarmac's *Flat roofing, a guide to good practice*, 1.1 Falls and drainage. Excellent explanation of needs to prevent ponding.

4 For avoiding restraint:

Note: None of the references given below, except Tarmac, give guidance for the use of bitumen-bound lightweight screeds. These need care in laying to avoid producing an uneven surface for the asphalt.

(a) CP 144: Part 4: 1970: *Roof coverings: mastic asphalt*, clause 3.6.2 for evenness of base to receive asphalt; clause 3.7 for isolating membrane and need for its continuity; and clause 3.11 (3) for cement-based screed finishes and need for separation by building paper, provision for expansion at bay joints (25 mm wide joints) and around perimeter (75 mm wide margin).

Note: It has been found that PVC or other plastics sheets do not provide adequate separation. The sheets are wrinkled by the wet screed and the wrinkles provide the restraint 2–3.

(b) PSA, *Technical guide to flat roofing*, 1987, 2.13.2 Movement with mastic asphalt on need for isolation in general and care to avoid use of slip layers and insulating boards that might offer some adhesive effect. Use of bitumen bound screeds not favoured (2.17.3). No advice on laid in-situ screeds.

(c) Tarmac's *Flat roofing, a guide to good practice*, 4.2 Asphalting techniques, for explanation of need for isolating layer (4.1 Materials, separating layer, page 116, gives advice on appropriate materials); 4.3 Surface protection, for isolating layer under all types of screeded finishes; 2.2 Screeds, for precautions with lightweight screeds, cement and bitumen bonded.

Manifestation 2

Minor cracking in asphalt or built-up bitumen felt, **7–9**.

Cause

1 In asphalt: loss of volatiles that usually results in the shrinkage of the asphalt.

2 In built-up bitumen felt: the combined effects of solar heating, ultraviolet radiation and atmospheric oxidation, commonly referred to as 'heat ageing' of the bitumen. Where there are chippings, the bitumen used to bond them to the surface of the felt cracks first, followed by the top layer of felt.

Remedial work

1 If the cracking is superficial and shallow (i.e. the cracks are less than 3 mm deep), apply a solar reflective treatment.

2 If the cracking is deeper than 3 mm but does not penetrate the membrane the application of one of the new polyurethane/coal: tar coatings such as Tretol's Roofguard and a solar reflective paint treatment may prolong the life of the membrane for up to five years.

3 If the cracks have penetrated the membrane but are localised, partial reinstatement is required. If the cracks have penetrated the membrane and are widespread on the roof, then complete reinstatement is required. Reroofing may be an alternative. If the membrane was BS 747 bitumen felt, consideration should be given to the use of one of the high-performance felts, in which case the manufacturer should be consulted for recommendations as there is no Code of Practice for these felts. BS 747 now includes (high-performance) polyester-based felts only.

References for correct detailing

1 For reinstatement of the membrane generally:

(a) CP 144: Part 3: 1970: *Roof coverings: built-up bitumen felt*.

(b) CP 144: Part 4: 1970: *Roof coverings: mastic asphalt*.

(c) BRE Digest 8, *Built-up felt roofs*. For minor remedial work.

(d) PSA, *Technical guide to flat roofing*, 1987, Sections 3.4–3.7.

(e) PSA, *Flat roofs, Technical Guide* (2nd edition), part 3. *Maintenance*. The other references above touch briefly on maintenance. This guide provides the clearest guidance on maintenance.

Note: The nature of the existing substrate will determine detailed considerations.

7 *An example of minor cracking: 'crocodiling' in felt. (Photo Building Research Establishment, Crown Copyright)*

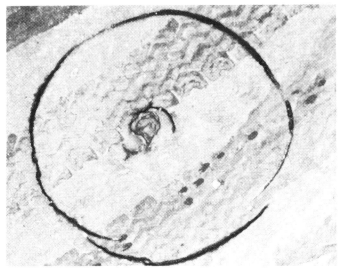

8, 9 *Minor cracking in asphalt, due to loss of volatiles. Distinguished from surface crazing in that crack penetrates full depth of waterproof membrane down to substrate. (Photos Building Research Establishment, Crown Copyright)*

(f) Tarmac's *Flat roofing, a guide to good practice*, Section 4, Mastic asphalt.

(g) BRC TIS 12 for polyester-based high-performance felts – basic information only.

2 For solar reflective treatments:

(a) CP 144: Part 4: 1970: *Roof coverings: mastic asphalt*, clause 3.11, suggests:

• Light-coloured chippings of limestone, granite, gravel, calcite or felspar 6–10 mm in size, set in bitumen dressing compound immediately after laying the asphalt.

• Concrete or asbestos cement tiles bedded in bitumen bonding compound.

• Cement/sand screed, 25 mm thick, cut into small squares (not more than 600 mm) with a 75 mm margin at the edges for movement. The edge margin should be filled with a bitumen compound.

Note: The CP advises against the use of aluminium-based solar reflective paints as permanent protection but see 2(c) below.

(b) PSA, *Technical guide to flat roofing*, 1987, 3.12 Protective layer, pp. 157–158 tabulates all types of protective layers, including those in CP 144 (a) above, but is biased towards the inverted system.

(c) GLC, *Development and Materials Bulletin*, **119**, October 1978, item 1, 'A durability trial of solar reflective paints on asphalt' and 120, December 1979, item 3, 'Solar reflective coatings on asphalt – their effectiveness'. Solar reflective paints will give a temperature reduction of between 15 and 35 per cent depending on the age of the coat and the amount of dust on it. The paint coating will not prevent blisters or slumping but will help to reduce the severity of these defects. Light-coloured chippings are the most effective treatment and where these are used reflective paints should be applied to the skirtings where the chippings cannot be laid.

Note: MACEF advises its members against the use of solar reflective paints due mainly to the need for continual maintenance of the coating.

Some aluminium-based solar reflective paints appear to cause much wrinkling of the surface of asphalt. Why this should be so has yet to be explained conclusively.

(d) Tarmac's *Flat roofing, a guide to good practice*, 4.3. Surface protection, pages 120 and 121, discusses the

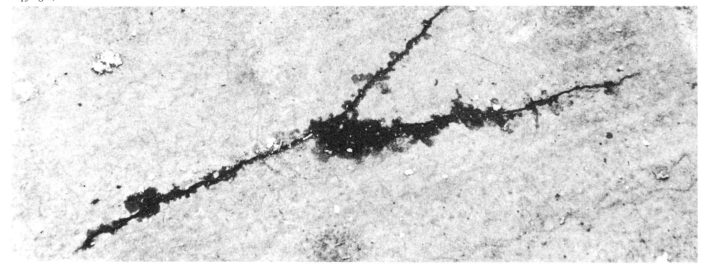

merits of different ways of achieving surface protection of asphalt.
3 For reroofing.
  (a) Tarmac's *Flat roofing, a guide to good practice,* 6.2 Reroofing. Excellent coverage of principles and details including choice of alternatives.

Manifestation 3
Major cracking in asphalt or built-up bitumen felt, **10–12**.

Cause
1 Differential movement between the waterproofing membrane and its substrate, arising from thermal and/or moisture movements; or
2 Movement of the structure due to inadequate provision of movement joints.

Cross references
Study 24: Movement, page 134.

Remedial work
In principle, a movement joint or joints need to be incorporated.

1 If the cracking is due to movement of the structure, it is unlikely that a movement joint incorporated to accommodate the differential movement between the membrane and its substrate will be effective for long. To deal with the movement of the structure, a movement joint extending through the whole of the structure (the roof and the walls) is required, but such a joint is likely to be extremely difficult to achieve in an existing structure.
2 If the cracking is due only to differential movement between the membrane and its substrate (the most likely cause where none of the dimensions of the roof exceed 30 m), then a movement joint should be incorporated. The joint may be either of the flush or upstand type. A flush joint is more easily achieved in remedial work. This type consists of making a chase in the substrate that is filled with a backing strip and sealant and then covered with a synthetic rubber strip bonded to the water proofing membrane. The sealant should be compatible with bitumen, such as Expandite's Plastiseal. Suitable rubber strips include Radflex and Expoband. An upstand type enables the joint to be covered with the vulnerable parts above the main level of the roof and is to be preferred.

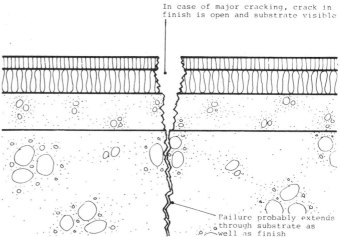

In case of major cracking, crack in finish is open and substrate visible

Failure probably extends through substrate as well as finish

10, 11 *Major cracking in asphalt: general and close-up views. Distinguished from minor cracking (previous manifestation) in that crack is open and substrate usually visible between separated edges. (Photo GLC.)*

12 *Crack probably penetrates substrate.*

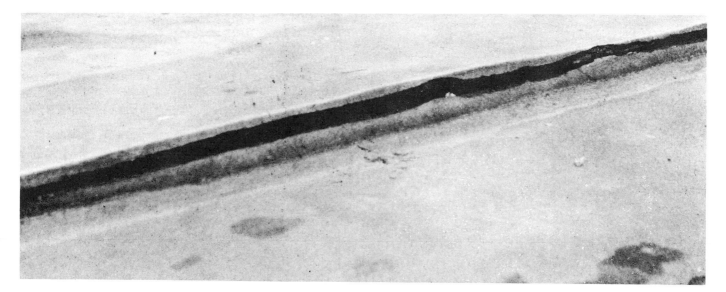

References for correct detailing
Movement joints
1 Flush types – for felt:
  (a) CP 144: Part 3: 1970: *Roof coverings: built-up bitumen felt*, page 37, Fig. 15.
  (b) BRE Digest 8 (new edition 1970), *Built-up felt roofs*, page 4, Fig. 3.
  (c) DOE, *Construction*, **30**, July 1979, 'Repairing cracks in felt roofing'. This emphasises the need for a slip strip in the vapour barrier as well as in the water proofing membrane itself.
  (d) PSA, *Technical guide to flat roofing*, 1987, 3.10, p. 151, Figure 3.172 for inverted roof only; p. 152, Figure 3.173 for joining existing to new inverted roof.
2 Flush types – for asphalt:
  (a) PSA, *Flat roofs, Technical Guide* (2nd edition), part 3, *Maintenance*, paragraph 319.4, page 73.
3 Upstand types – for felt:
  (a) CP 144: Part 3: 1970: *Roof coverings: built-up bitumen felt*, page 37, Fig. 17.
  (b) PSA *Technical guide to flat roofing*, 3.10.13, Figures 3.167 to 3.171, p. 149–151, include a range of decks and forms of joint.
  (c) FRCAB *Roofing handbook*, Information sheet 10, last page.
4 Upstand types – for asphalt:
  (a) CP 144: Part 4: 1970: *Roof coverings: mastic asphalt*, page 41, Fig. 15.
  (b) MACEF, *Roofing handbook*, page 26. Gives two clear figures.
  (c) PSA, *Technical guide to flat roofing*, 1987, 3.6.13, p. 65, Figure 3.71 for inverted roof.
  Note: All the references emphasise the need for providing some form of solar reflective treatment if none exists already. (See References for correct detailing for Manifestation 1.)

13, 14 *Ridges in built-up felt over gaps and unevenness in the insulation boards below. Distinguished from the following two manifestations by the fact that the ridges are firm when pressed rather than yielding. In* **9** *the substrate was expanded polystyrene;* **10** *shows conditions with strawboard (foreground) and tongue-and-groove boarding (behind). Note the effect on ponding.*

Manifestation 4

Ridging of built-up bitumen felt. The ridges are firm when pressed, **13**, **14**. (For ridges which yield when pressed, see Manifestations 5 and 6.)

Cause
1 The substrate – such as timber boards, woodwool slabs, chipboards, flax boards and insulation boards – was uneven before the felt was laid or has distorted due to moisture movement during service, or,
2 Thermal and/or moisture movements of the substrate with or without a vaporisation of moisture entrapped below the waterproofing membrane. The ridging always occurs over or near the joints between the boards. Vaporisation of entrapped moisture usually inflates the ridges.

Remedial work
If there are leaks or if the ridges seriously interfere with the roof drainage:

1 If the substrate is uneven or has distorted, strip off the felt and substrate and relay entirely.
2 If the substrate is even and has not distorted but is damp, strip off the felt and relay entirely installing large type drying units at low and high points in the roof, spaced according to the manufacturer's recommendations and with evaporation tubes inserted in the units.
3 If the substrate is even and has not distorted and thermal movements have been responsible for the ridging (notable with expanded plastics with high coefficients of thermal expansion), strip off the felt and relay entirely with new felt on 13 mm fibre board bonded with hot bitumen to the substrate. Consideration should be given to the use of high-performance felts, in which case the manufacturers should be consulted for his recommendations as there is no Code of Practice for this type of felt. Polyester-based high-performance felts only are now included in BS 747.

References for correct detailing
1 CP 144: Part 3: 1970: *Roof coverings: built-up bitumen felt*, clause 2.4 'Roof deck materials' divides substrates into two groups. The first includes: concrete, metal decking, asbestos cement decking, woodwool slabs and treated timber and plywoods. The second group is prefaced by a warning that under adverse conditions there is a high risk of movement and includes: untreated timber, chipboard, blockboard and laminboard.
2 BRE Digest 8: *Built-up felt roofs*, Table 1, page 5, lists the range of substrates, the treatment before laying the felt and the correct method of fixing the first layer.
3 PSA, *Technical guide to flat roofing*, 1987, 3.10, p. 119–152 deals in detail with all type of roof deck.
4 BFRC TIS 12 for polyester-based high-performance felts – basic information only.

15 *Undulations in built-up felt. Differs from ridging in that the raised profile yields to pressure easily.*

Manifestation 5
Undulations in built-up roofing felt which yield to pressure, **15** sometimes in conjunction with ridging, **17**.

Cause
1　Felt distorted when laid because rolls were stored on sides, not on ends, and
2　Inadequate pressure during laying, or,
3　Insufficient or badly distributed bitumen compound, and/or
4　The vaporisation of entrapped moisture in the roof deck.

Cross reference
Study 21: Entrapped moisture, page 114

Remedial work
1　Generally none is required – the application of a solar reflective treatment will help to reduce further inflation – unless the undulations are likely to be damaged by maintenance and other traffic on the roof, in which case strip off the existing and relay with new felt.
2　Where entrapped moisture is the main cause of the problem – some dampness is likely to have been found below the felt – it is essential that adequate protection, such as temporary roofing, is used to prevent rainwater becoming entrapped during remedial work. As a further precaution, ventilators should be installed to allow any further drying out to take place. Large type drying units with evaporation tubes are to be preferred – install according to the manufacturer's recommendations.

References for correct detailing
1　For relaying felt generally:
　　Note: None of the references given below deal with high-performance felts. These are particularly useful where the roof is being upgraded thermally and expanded plastics boards are used. Many of these boards have large thermal movements which the new felts can better accommodate. Consult the manufacturer for details of the use of his felt.
　(a)　CP 144: Part 3: 1970: *Roof coverings: built-up bitumen felt*, clause 2.4 'Roof deck materials' divides substrates into two groups. The first includes: concrete, metal decking, asbestos cement decking, woodwool slabs and treated timber and plywoods. The second

group is prefaced by a warning that under adverse conditions there is a high risk of movement and includes: untreated timber, chipboard, blockboard and laminboard.
　(b)　BRE Digest 8: *Built-up felt roofs*, Table 1, page 5, lists the range of substrates, the treatment before laying the felt and the correct method of fixing the first layer.
　(c)　PSA, *Flat roofs, Technical Guide* (2nd edition) pages 26–34 deals in detail with built-up felt on lightweight decks. Clauses 139–141 deal with timber, woodwool and metal trough decks. In particular schedule 145.5 on page 26 lists the range of decks as BRE Digest 8 but in more detail.
　(d)　Felt Roofing Contractors Advisory Board Guide, *Built-up roofing*. The illustrations are very diagrammatic but the principles are shown clearly.
2　For solar reflective treatments:
　(a)　CP 144: Part 4: 1970: *Roof coverings: mastic asphalt*, clause 2.9 suggests:
　　●　Light-coloured chippings of limestone, granite, gravel, calcite or felspar 6–10 mm in size, set in bitumen dressing compound immediately after laying the asphalt.
　　●　Concrete or asbestos cement tiles bedded in bitumen bonding compound.
　　●　Cement/sand screed, 25 mm thick, cut into small squares (not more than 600 mm) with a 75 mm margin at the edges for movement. The edge margin should be filled with a bitumen compound.

Note: The CP advises against the use of aluminium-based solar reflective paints as permanent protection but see 2(c) below.
　(b)　PSA, *Flat roofs, Technical Guide* (2nd edition) clause 124, page 19 recommends the use of chippings (14 mm size) embedded in bitumen to BS 3690, grade 25s at the rate of 15 kg/m^2.
　(c)　GLC, *Development and Materials Bulletin*, **119**, October 1978, item 1, 'A durability trial of solar reflective paints on asphalt' and 120, December 1979, item 3, 'Solar reflective coatings on asphalt – their effectiveness.' Solar reflective paints will give a temperature reduction of between 15 and 35 per cent, depending on the age of the coat and the amount of dust on it. The pain coating will not prevent blisters or slumping but will help to reduce the severity of these defects. Light-coloured chippings are the most effective treatment, and where these are used reflective paints should be applied to the skirtings where the chippings cannot be laid.
　　Note: MACEF advises its members against the use of solar reflective paints due mainly to the need for continual maintenance of the coating.
　　Some aluminium-based solar reflective paints appear to cause much wrinkling of the surface of asphalt. Why this should be so has yet to be explained conclusively.
　(d)　Tarmac's *Flat roofing, a guide to good practice*, 4.3. Surface protection, pages 120 and 121, discusses the merits of different ways of achieving surface protection of asphalt.
3　For reroofing.
　(a)　Tarmac's *Flat roofing, a guide to good practice*, 6.2 Reroofing. Excellent coverage of principles and details including choice of alternatives.

Manifestation 6

Blistering of built-up roofing felt – the blisters yield to pressure, **16** sometimes in conjunction with ridging.

Cause

Inadequate pressure during laying causing air pockets that subsequently 'absorb' moisture from the atmosphere or, more seriously, vaporisation of moisture trapped, either below the roof covering (source of moisture: construction water and/or condensation) or between the layers of felt.

Cross reference

Study 21: Entrapped moisture, page 114

Remedial work

1 *At best*: If there are no leaks:
  (a) Apply solar reflective treatment, or,
  (b) Star cut the blister, rebound to the under layer or base and then patch over the affected area. Apply a solar reflective treatment. If there is entrapped moisture this must be released (e.g. by installing large type drying units with evaporation tubes inserted and the units spaced according to the manufacturer's recommendations).
2 *At worst*: If there are leaks (but this is unlikely) replace the felt. Any entrapped moisture must be allowed to dry out first. If the moisture is due to interstitial condensation

(see Studies 7 and 8 for diagnosis) install large type drying units with evaporation tubes inserted – the units to be spaced according to the manufacturer's recommendations.

Reference for correct detailing

1 For relaying felt generally:
  None of the references given below deal with high-performance felts. These are particularly useful where the roof is being upgraded thermally and expanded plastics boards are used. Many of these boards have large thermal movements which the high-performance felts can better accommodate. Consult the manufacturer for details of the use of his felt.
  (a) CP 144: Part 3: 1970 (including AMD 2527, April 1978): *Roof coverings: built-up bitumen felt*, clause 2.4 'Roof deck materials', divides substrates into two groups. The first includes: concrete, metal decking, asbestos cement decking, woodwool slabs and treated timber and plywoods. The second group is prefaced by a warning that under adverse conditions there is a high risk of movement and includes: untreated timber, chipboard, blockboard and laminboard.
  (b) BRE Digest 8: *Built-up felt roofs*, Table 1, page 5, lists the range of substrates, the treatment before laying the felt and the correct method of fixing the first layer.

16 *Blistering of built-up felt, similar to undulations in that the raised areas will yield to pressure but different in that they are distributed as 'islands' rather than continuous corrugations.*

17 *Blistering, ridging, birdsmouthing, poor lap adhesion and ponding. Two layers of organic fibre ('rag') felt on fibreboard.*

(c) PSA, *Technical guide to flat roofing*, 1987, 3.10, pp. 119–152 deals in detail with all types of roof deck.

(d) FRCAB *Roofing handbook,* June 1988, Information sheets 9 and 10.

2 For solar reflective treatments:

(a) CP 144: Part 4: 1970: *Roof coverings: mastic asphalt,* clause 2.9, suggests:

- Light-coloured chippings of limestone, granite, gravel, calcite or felspar 10 mm in size, set in bitumen dressing compound immediately after laying the asphalt.
- Concrete or asbestos cement tiles bedded in bitumen bonding compound.
- Cement/sand screed, 25 mm thick, cut into small squares (not more than 600 mm) with a 75 mm margin at the edges for movement. The edge margin should be filled with a bitumen compound.

The CP makes no recommendations for the use of solar reflective paints but see 2(c) below.

(b) PSA, *Technical guide to flat roofing,* 1987, 3.12 Protective layer, pp. 157–158 tabulates all types of protective layers, including those in CP 144 (a) above, but is biased towards the inverted system.

(c) GLC, *Development and Materials Bulletin,* **119**, October 1978, item 1, 'A durability trial of solar reflective paints on asphalt' and **120**, December 1979, item 3, 'Solar reflective coatings on asphalt – their effectiveness'. Solar reflective paints will give a temperature reduction of between 15 and 35 per cent, depending on the age of the coat and the amount of dust on it. The paint coating will not prevent blisters or slumping but will help to reduce the severity of these defects. Light-coloured chippings are the most effective treatment and where these are used reflective paints should be applied to the skirtings where the chippings cannot be laid.

Note: MACEF advises its members against the use of solar reflective paints due mainly to the need for continual maintenance of the coating.

Some aluminium-based solar reflective paints appear to cause much wrinkling of the surface of asphalt **4**. Why this should be so has yet to be explained conclusively.

(d) Tarmac's *Flat roofing, a guide to good practice,* 4.3. Surface protection, pages 120 and 121, discusses the merits of different ways of achieving surface protection of asphalt.

3 For reroofing:

(a) Tarmac's *Flat roofing, a guide to good practice,* 6.2 Reroofing. Excellent coverage of principles and details including choice of alternatives.

4 For ventilators:

Manufacturer's literature (e.g. D. Anderson and Son Limited and Briggs Amasco Limited). Consult the latest edition of *Specification.*

Manifestation 7
Blistering of asphalt.

Cause
Vaporisation of moisture or air entrapped in laying – often between the two coats of asphalt – or during construction (rainwater, for example) or during service (interstitial condensation).

Cross references
Study 21: Entrapped moisture, page 114.

Remedial work
1 *At best*: if the blisters are not split and there are no leaks apply a solar reflective treatment. As a precaution, entrapped moisture should be released (e.g. by inserting large type drying units with evaporation tubes inserted, the units spaced according to the manufacturer's recommendations).
2 *At worst*: if there are leaks or if the blisters are split, remove the blisters and repair with asphalt. Treat any entrapped moisture as in 3 above.

References for correct detailing
For repairs to asphalt: CP 144: Part 4: 1970: *Roof coverings: mastic asphalt,* clause 5.2, page 27, describes the method used for repairing existing asphalt. New hot asphalt is placed around the blister until it is sufficiently softened to be cut out without causing cracking. When the new asphalt is laid, the junction between the old and new material should be made with a lapped joint so that complete fusion takes place.

Manifestation 8: Splitting, tearing or debonding of single-layer membranes

Condition
Single-layer (or plies, as they are also described) membranes are being used increasingly. Defects in them are therefore emerging. There are different types of membrane, depending on the chemical type used, e.g.:

- PVC – Polyvinyl chloride
- CPE – Chlorinated polyethylene
- CSM – Chlorosulphonated polyethylene
- EPDM – Ethylene propylene diene terpolymer
- PIB – Polyisobutylene

Some are laid loose on the substrate; some are bonded and/or mechanically fastened to the substrate; all have bonded laps or seams. Some have Agrément Certificates.

The causes given below are therefore of a general, if not preliminary, nature.

Cause
One or more of:

1 Laid on sharp objects such as pieces of grit, stone or debris and then subject to foot traffic during laying or subsequently.
2 Absence of a separating layer between the membrane and the underlying substrate which is of a chemically incompatible material (bitumen, for example, 'attacks' some of the membranes).
3 Lack of the exceptionally good workmanship (cleanliness no less before and during bonding) required at laps, corners and fixings.
4 Puncturing by fixings (for PVC membranes in particular) due to intermittent wind-uplift.
5 Migration of the plasticiser.

Cross references
Study 25: Loss of adhesion for principles only. Chemical properties not covered in any studies but are in *Materials for*

building, Vol. 1: 1.08 The basic structure of matter, pages 49–53; 1.09 Chemical conventions, pages 54 and 55; and 1.10 Chemical reactions (includes discussion of different types of plastics), pages 56–59.

Remedial work
At best: Local repair of the membrane – most are easier to patch than asphalt or bitumen felt.

At worst: Strip and reinstate.

References for correct detailing
1 Agrément Certificates (nine currently; others applied for).
2 Individual manufacturers or through Single Ply Roofing Association (SPRA), 17 St George Street, London W1R 9DE.

Study 7
Flat roofs: concrete slab

Manifestation, Conditions 1 and 2

1 Internally: leakage of water from the underside of the slab usually at weak points such as cracks in the slab or electrical conduit outlets. The water may be a brownish colour.
2 Externally: rippling, tearing or blistering of the waterproof membrane that may or may not contribute to the water leakage. (See Study 6, Manifestations 4–7, pages 34–37.

Cause, Condition 1
The water leakage may be due to either one or all of:

1 The entrapment of water during construction, the water being derived from either rainfall or the construction process.
2 The entrapment of water absorbed by the dry insulation during screed laying. Screeds should never be laid on dry insulation. The moisture from the screed makes the dry insulation less effective, thereby increasing the risk of interstitial condensation.
3 Interstitial condensation.
4 Rainwater penetration through splits or tears in the membrane. Care should be taken in diagnosing rainwater penetration because the leaking water may have taken a long route from its point of entry.

If there is no rainwater penetration, leakage at the underside of the slab signifies the release of entrapped water and/or interstitial condensation. Brown staining of the leaching water signifies leaching of bitumen from roofing felts or bitumen felt underlay under the asphalt.

Rippling and tearing of the waterproof membrane (asphalt especially) is due probably to movement of the unrestrained screed (lightweight screeds notably).

Blistering of the waterproof membrane is due probably to the vaporisation of moisture (entrapped or from interstitial condensation) below it on exposure to solar radiation.

1 *Penetration of rainwater through roofslab at conduit hole. (Photo Building Research Establishment, Crown Copyright)*

2 *Leakage of entrapped moisture.*

3 *Concrete slab construction, without vapour control layer (Condition 1).*

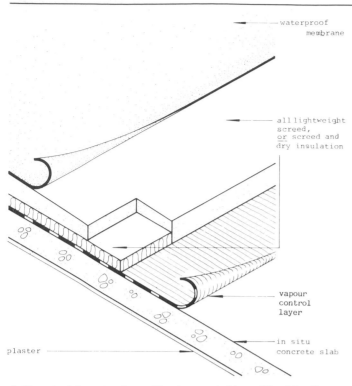

waterproof
membrane

all lightweight
screed,
or screed and
dry insulation

vapour
control
layer

in situ
concrete slab

plaster

4 *Concrete slab construction, with vapour control layer (Condition 2).*

Generally, the absence of a solar reflective treatment to the exposed surface of the waterproof membrane increases the rippling, ridging or blistering.

Cause, Condition 2
Generally the reasons for water leakage are the same as those described in Condition 1 with the exception of interstitial condensation, unless (and this is most important) the vapour control layer is ineffective because it lacks integrity (i.e. the vapour control layer has tears, faulty laps and is pierced by electrical and other services) or because there is insufficient thermal insulation above it. Rippling, tearing and blistering as described in Condition 1 above are the chief forms of failure.

Cross references
1 Study 20: Condensation. See under 'Interstitial condensation', 4.03, page 108.
2 Study 21: Entrapped moisture. See 'Diagnostic checklist', 5, page 116.
3 Study 22: Rain penetration. See 'Flat roofs', 6, page 118.
4 Study 25: Loss of adhesion. See 'Built-up bitumen felt to flat roofs', 11, page 154.

Diagnosing the cause
Identification of the source of moisture that is causing the water leakage – and such identification is most important to determine appropriate remedial work – must usually be done by a process of elimination. A reliable history of the times when the water leakage occurs is a valuable starting point. If not available, site tests and/or opening up is inevitable.

If the leakage occurs only during cold weather and is worse in sudden very cold spells, interstitial condensation is almost certainly the only or primary cause. If leakage occurs only during rain or shortly afterwards, the leakage is almost certainly due to rain penetration of the membrane or associated features. Interstitial condensation and rainwater would be responsible if the leakage occurs during rain and in cold weather.

Site testing for rainwater penetration should start with a careful systematic inspection of the surface of the roof including associated features. If such an inspection does not reveal the defect(s), water tests of the roof should be carried out. If these tests prove negative, then it is necessary to open up parts of the roof.

Full details of site testing with water are given in PSA, *Flat roofs, Technical Guide* (2nd edition), Annex E. pp. 90 and 91.

Note: Tests with fluorescein dyes use an ultraviolet light and need care in use. In the absence of authoritative guidance on their use, consult building testing laboratories.

Remedial work: condensation
If interstitial condensation has been identified as the only cause of water leakage and the waterproof membrane is not damaged anywhere, it is worthwhile investigating the installation of a protected membrane roof (also known as an 'inverted roof', 'upside down roof' and an 'inside out roof'). In principle, this means laying new thermal insulation with a low water absorption (e.g. an extruded polystyrene board such as Dow's Roofmate) so as to keep the roof construction warm and the underside of the waterproof membrane – that in effect is also acting as a vapour control layer – above the dewpoint. The thermal insulation board has to be held down against wind up by either pre-paved boards or 50 mm gravel or 50 mm precast concrete slabs. The exposed vertical surfaces should be treated with a solar reflective paint. The roof slab may, however, not be strong enough to support the weight of the pre-paved surface or the gravel of the concrete slabs. If this is the case, plug all the gaps around light drops and other perforations in the vapour control layer with a sealant, add a vapour control layer to the face of the ceilings (for example, PVC faced/foil backed paper or rubber paint) if moisture vapour from the interior is a contributory factor, and/or install drying units (as described for 'Entrapped moisture' Study 21).

References for correct detailing
1 General:
   (a) BRE Digest 312, *Flat roof design: the technical options*, gives comprehensive guidance on the principles of warm and cold deck roof designs.
   (b) PSA, *Inverted roofs, Technical Guide* in association with Bickerdike Allen Partners (1984), clear and concise. But see PSA *Technical guide to flat roofing*, 1987, for new rules as to 200 mm min skirting heights.
   (c) DOE, *Construction* **31**, October 1979, pp. 20–21 describes comparative tests by the PSA with a conventional roof. The temperature fluctuations in the asphalt in the inverted roof are shown to be much less than in a conventional roof, which means the risk of thermal movement and consequent cracking is reduced. The rate of temperature change is less, which again reduces the risk of cracking.
   (d) DOE, *Construction* **14**, page 46 and **17**, page 47, gives details of rainwater outlets for use with inverted roofs.
   (e) GLC, *Development and Materials Bulletin* **90**, item 2, page 4, gives a good summary of the argument for using inverted roofs:
     ● The membrane is protected from solar degradation and mechanical damage.
     ● The membrane is protected from temperature extremes, yet remains accessible for maintenance.

- As there is only one vapour control layer (the waterproofing membrane acting as a vapour control layer as well) construction moisture can dry out and does not become entrapped.

(f) GLC, *Development and Materials, Bulletin* **147**, item 5, 'Experimental inverted roof systems' describes the result of success in the experiment. A follow-on from *Bulletin* **90**, reference (d) above.

2 For ventilators:

(a) PSA, *Technical guide to flat roofing*, 1987, 2.10.16, Figure 2.22, page 43.

(b) *Tarmac's Flat roofing, a guide to good practice*, 1.2, pages 17 and 18.

(c) Manufacturer's literature (e.g. D. Anderson & Son Ltd., Briggs Amasco Ltd., Screeduct Ltd., Euroroof Ltd.). Consult the latest *Specification*.

3 For applying vapour control layers:

Manufacturer's literature for

- Chlorinated rubber paint.
- PVC faced/foil backed paper.
- Gyproc vapour check wallboard (British Gypsum White Book). This can be upgraded to a vapour control layer quality by sealing the joints between the boards with a vapour resistant tape (e.g. Densotape made by Winn and Coles Limited or with a gun applied sealant compatible with the polyethylene film backing).
- Plastapac made by Acalor International of Crawley is a highly effective but costly spray-on barrier. Acalor apply their material themselves and do not normally recommend Plastapac in domestic buildings.

Note: The amount of water vapour transmitted through cracks and gaps around light drops and services, is significant – proportionately far greater than would be expected from the size of the split or gap. Apart from ensuring that all joints are securely lapped (preferably bonded with an adhesive) or securely taped, all gaps around light drops and other services passing through the vapour control layer should be plugged with a sealant. To reduce the risk of water vapour bypassing the vapour control layer at the wall/ceiling junction, the vapour control layer should be taken down the wall for a distance of at least 200 mm.

Remedial work: entrapped moisture

If entrapped moisture has been identified as the main cause of leakage or if it has caused splitting of blisters and ripples, then some means of deck and screed ventilation must be installed. The means adopted to provide adequate ventilation depends on the amount of moisture that is entrapped and the extent to which blisters and ripples have split.

At best: Install large type drying units incorporating evaporation tubes at low and high points in the roof, spaced in accordance with the manufacturer's recommendations. Consideration might also be given to pumping water out of the screed or draining water out of the screed by drilling through the soffit, taking care to have a stop on the drill so as not to damage the waterproofing membrane.

At worst: Remove the waterproofing membrane, allow the screed to dry out and lay a new waterproof membrane. It is essential that adequate protection (such as a temporary roof) is used to prevent rainwater becoming entrapped during the remedial work. If the drying out of the roof cannot be achieved or if further drying out is required, install a screed duct system such as Screedmat, 'a screed ventilation system laid over the existing screed with appropriately located ventilator cowls', and then lay the waterproof membrane. Some form of solar reflective treatment should be applied in all cases.

References for correct detailing

1 For ventilators:

(a) Manufacturer's literature (e.g. D. Anderson and Son Limited and Briggs Amasco Limited). Consult the latest edition of *Specification*.

(b) Screedmat made by Screeduct Limited of 109 Northwood Road, Thornton Heath, Surrey, is a screed ventilation system used successfully by the GLC (see GLC, *Development and Materials Bulletin* **92**, item 2, page 4, February 1976).

2 For replacing the membrane generally – built-up felt:

(a) CP 144: Part 3: 1970: *Roof coverings: built-up bitumen felt*.

(b) BRE.Digest 8: *Built-up felt roofs*, 1970.

(c) PSA, *Technical guide to flat roofing*, 1987, 3.10, pp. 119–152 deals in detail with all type of roof deck.

(d) FRCAB *Roofing handbook*, June 1988, Information sheets 9 and 10.

(e) Tarmac's *Flat roofing, a guide to good practice*, 'Built-up roofing', 3.2 'Application techniques' (pages 71–74).

3 For replacing the membrane generally – asphalt:

(a) CP 144: Part 4: 1970: *Roof coverings: mastic asphalt*.

(b) PSA, *Technical guide to flat roofing*, 1987, 3.6, pp. 29–67.

(c) MACEF, *Roofing handbook*. The detailed drawings are extremely clear.

4 For solar reflective treatments:

(a) CP 144: Part 4: 1970: *Roof coverings: mastic asphalt*, clause 2.9 suggests:

- Light-coloured chippings of limestone, granite, gravel, calcite or felspar 6–10 mm in size, set in bitumen dressing compound immediately after laying the asphalt.
- Concrete or asbestos cement tiles bedded in bitumen bonding compound.
- Cement/sand screed, 25 mm thick, cut into small squares (not more than 600 mm) with a 75 mm margin at the edges for movement. The edge margin should be filled with a bitumen compound.

Note: The CP advises against the use of aluminium-based solar reflective paints as permanent protection.

(b) PSA, *Technical guide to flat roofing*, 1987, 3.12 Protective layer, pp. 157–158 tabulates all types of protective layers, including those in CP 144 (a) above, but is biased towards the inverted system.

(c) GLC, *Development and Materials Bulletin* **119**, October 1978, item 1, 'A durability trial of solar reflective paints on asphalt' and 120, December 1979, item 3, 'Solar reflective coatings on asphalt – their effectiveness'. Solar reflective paints will give a temperature reduction of between 15 and 35 per cent, depending on the age of the coat and the amount of dust on it. The paint coating will not prevent blisters or slumping but will help to reduce the severity of these defects. Light-coloured chippings are the most effective treatment and where these are used reflective paints should be applied to the skirtings where the chippings cannot be laid.

Note: MACEF advises its members against the use of solar reflective paints due mainly to the need for continual maintenance of the coating.

Some aluminium-based solar reflective paints appear to cause much wrinkling of the surface of asphalt. Why this should be so has yet to be explained conclusively.

(d) Tarmac's *Flat roofing, a guide to good practice*, 4.3. Surface protection, pages 120 and 121, discusses the merits of different ways of achieving surface protection of asphalt.

5 For reroofing:

(a) Tarmac's *Flat roofing, a guide to good practice*, 6.2 Reroofing. Excellent coverage of principles and details including choice of alternatives.

Remedial work: movement

If rippling, tearing or cracking has been due to movement of the substrate under the waterproof membrane and the substrate is dry:

1 *At best*: Repair of the affected parts by a specialist sub-contractor may be possible.

2 *At worst*: Remove the waterproof membrane and relay either with isolating strips over minor cracks in the screed or with movement joints over major cracks. The latter may be either of the flush or the upstand type. A flush joint is more easily achieved in remedial work. This type consists of making a chase in the substrate that is filled with a backing strip and sealant and then covered with a synthetic rubber strip bonded to the waterproofing membrane. The sealant should be compatible with bitumen, such as Expandite's Plastiseal. Suitable rubber strips include Radflex and Expoband. An upstand type enables the joint to be covered with the vulnerable parts above the main level of the roof and is to be preferred.

References for correct detailing: movement joints

1 Flush types – for felt:

(a) CP 144: Part 3: 1970: *Roof coverings: built-up bitumen felt*, page 37, Fig. 15

(b) BRE Digest 8 (new edition 1970), *Built-up felt roofs*, page 4, Fig. 3.

(c) DOE, *Construction* **30**, July 1979, 'Repairing cracks in felt roofing'. This emphasises the need for a slip strip in the vapour control layer as well as in the waterproofing membrane itself.

(d) PSA, *Technical guide to flat roofing*, 1987, 3.10, pages 119–152 deals in detail with all type of roof deck.

2 Flush types – for asphalt:

(a) PSA, *Flat roofs, Technical Guide* (2nd edition) paragraph 319.4, page 73.

3 Upstand types – for felt:

(a) CP 144: Part 3: 1970: *Roof coverings: built-up bitumen felt*, page 37, Fig. 17.

(b) Tarmac's *Flat roofing, a guide to good practice*, pages 158 and 168.

(c) FRCAB *Roofing handbook*, June 1988, Information sheets 9 & 10.

4 Upstand types – for asphalt:

(a) CP 144: Part 4: *Roof coverings: mastic asphalt*, page 41, Fig. 15.

(b) Mastic Asphalt Council: *Roofing handbook*, page 26. Gives two clear figures.

(c) PSA, *Technical guide to flat roofing*, 1987, 3.6.13, p. 65, Figure 3.71 for inverted roof.

Note: All the references emphasise the need for providing some form of solar reflective treatment if none exists already. (See References for correct detailing under Remedial work: Entrapped moisture.)

Study 8
Flat roofs: timber

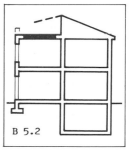

B 5.2

Manifestation

In the absence of rain, and particularly during cold weather, there is damp staining of the ceiling and/or decay of the timber joists or decay of the decking that may be of timber (e.g. boarding), timber products (e.g. plywood, blockboard, laminboard, woodwool slabs and chipboard) or other organic materials (flaxboard and strawboard), and/or mould growth within the construction.

If there is decay and/or mould growth there is likely to be a musty smell; cracked ceilings, excessive ponding on the roof and/or defects in the waterproof membrane such as splitting, tearing or (in the case of built-up felts) the opening up of laps may indicate that the decay of the joints and/or decking is in an advanced state and that structural weakening has occurred.

Cause

In all cases the basic cause is interstitial condensation that in turn is due to the causes given for each condition described below. Rainwater penetration may be a secondary or contributory cause if the waterproof membrane is torn or split due to structural weakening and/or moisture movement of the roof deck.

Note: In those cases where the timber has not resisted decay due to the damp conditions arising from the interstitial condensation the timber could not have been treated with a timber preservative.

Remedial work

In all cases, consideration should be given first to the extent to which the moisture content of the air in contact with the

1 *Mould growth on ceiling. Diffuseness of damp manifestation (as opposed to well-defined patches) and pressure of mould (which can only grow in air of relative humidity not les than 70 per cent) suggest condensation as probable major cause. (Photo GLC.)*

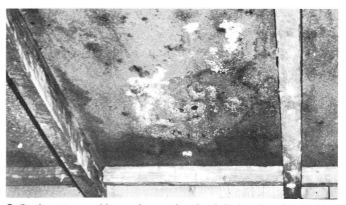

2 *Condensation mould growth on underside of chipboard roof deck. Mould manifested as grey, green, black or brown spots or patches, which may spread to become a furry layer. (Photo Building Research Establishment, Crown Copyright)*

3 *Furry mould growth on roof timbers; probably associated with condensation for reasons given under* **1**. *(Photo Building Research Establishment, Crown Copyright)*

43

ceiling can be reduced by better ventilation. (In warm deck designs the vapour control layer may not be necessary if adequate ventilation is provided – see References for correct detailing.) Increased ventilation may require an increase in heating.

Wherever possible, existing unpreserved timber that is to remain in position should be given an *in situ* preservative treatment. Any new timber used must be treated with a preservative, preferably pressure impregnated.

Cross references
Study 20: Condensation. See under 'Interstitial condensation', **4.03**, page 108. *Note*: See, in particular, replacement of 'vapour barrier' and 'vapour check' by 'vapour control layer'. The latter is used in this study.
Study 27: Timber decay, page 162.

Condition 1: Warm deck roof (1)

Cause
1 There is no vapour control layer under the insulation.
2 The cavity has been ventilated.

Note: In this type of roof the cavity should not be ventilated if there is a vapour control layer because it reduces the thermal insulation. But in the absence of a vapour control layer, ventilation of the cavity may have helped to control the condensation.

Remedial work
1 *At best*:
(a) If the cavity is ventilated, seal all ventilation openings to the cavity – and any other openings that may be providing ventilation inadvertently – and consider (b) below.
(b) If the structure and deck have not decayed (or have only decayed very slightly) and if the waterproof membrane has not been badly torn or split, use the existing waterproof membrane as a vapour control layer and add new thermal insulation and a

waterproof membrane on top either conventionally or with one of the spray-on polyurethane (or similar) foam/waterproofing compounds offered by specialists such as Ruberoid (see DAS 59 for precautions, whatever method is used). (The use of an inverted roof is usually not possible on timber roofs because of the considerable additional load imposed by the gravel or concrete slabs that must be laid on top of the thermal insulation.)
All chippings embedded in bitumen must be removed as thoroughly as is possible. The depth of thermal insulation required may be thick enough to affect other details such as the height of skirtings and rainwater outlets. These should be given special consideration.
2 *At worst*:
(a) Strip off the waterproof membrane and insulation. Replace with an appropriate vapour control layer on top of the roof deck followed by new thermal insulation and a waterproof membrane.
(b) If, in addition, there has been decay of timber or other organic materials, the affected parts will have to be replaced. In an extreme case, this may well mean replacing the whole of the roof.

Condition 2: Warm deck roof (2)

Cause
1 If the insulation is wet, the vapour control layer:
(a) Has inadequate vapour resistance in itself for the vapour conditions within the space and/or
(b) Lacks integrity because of breaks or splits in it and/or because of inadequate detailing at the edges (e.g. the vapour control layer was not lapped over the insulation) and because of inadequate detailing around openings for services (e.g. the gap was not filled with a sealant).
2 If the roof deck is wet and/or there is decay, it is almost certain that there is insufficient thermal insulation above

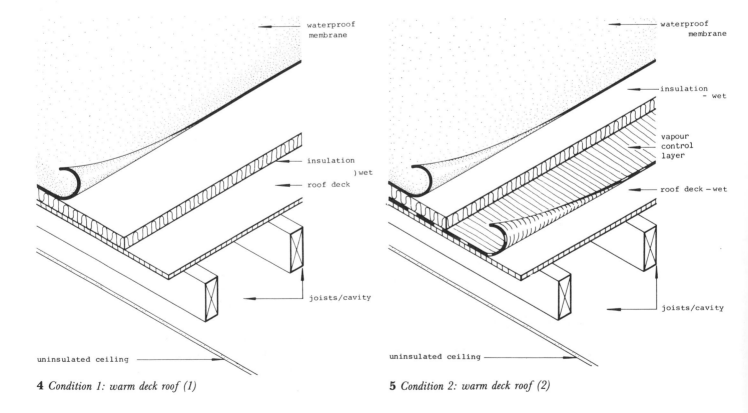

4 *Condition 1: warm deck roof (1)*   **5** *Condition 2: warm deck roof (2)*

the vapour control layer. The insufficiency of thermal insulation may be due to the ceiling itself providing too great a proportion of the thermal insulation of the roof.

Note: The ventilation of the cavity could mitigate against the successful performance of both the vapour control layer and the insulation described in 1 and 2 above.

Remedial work
1 *At best*:
Seal all ventilation openings to the cavity – and any other openings that may be providing ventilation inadvertently.
2 *At worst*:
(a) If the thermal insulation is inadequate, strip off the waterproofing membrane, add more thermal insulation and then lay a new waterproof covering. (Consult the references for predicting the sufficiency of the thermal insulation.)

Note: The removal of built-up felt may make it difficult not to damage the insulation unduly. If the damage to the insulation is likely to be significant, then it will be necessary to seriously consider overlaying the whole of the roof with new insulation as described in the remedial work for Condition 1: Warm deck roof (1), 1(b) above.
(b) If the vapour control layer is inadequate, strip off all the layers down to the roof deck. Lay a new vapour control layer, thermal insulation and waterproof membrane. A decision as to the adequacy of the vapour control layer will probably have to be made on the basis of the inspection of selected areas after opening up (great care is required in the opening up to avoid damaging the vapour control layer) and a survey of the openings in it made by light drops, electrical conduits and other service outlets.

Condition 3: Cold deck roof (1)

Cause
1 The cavity is inadequately ventilated.
2 The vapour control layer is ineffective or absent.

Note: Because of the difficulties of sealing the vapour control layer, inadequate ventilation is usually the most likely cause.

Remedial work
1 *At best*:
Increase the cavity ventilation so that the ventilation openings are on opposite sides of the roof and are equivalent to *at least* 0.4 per cent of the plan area – at least 0.6 per cent of the plan area for a low building or a building in a sheltered location. Where there are cavity barriers in the roof, as required by Building Regulations, cowl ventilators will have to be installed to achieve adequate ventilation of the isolated zones.
2 *At worst*:
In addition to increasing the cavity ventilation, apply a vapour control layer to the underside of the ceiling making sure that the gaps around any perforations through the ceiling are properly filled with a sealant. The vapour resistance required will depend on circumstances. BS 5250 and BS 6229 give basic data on the vapour resistance of suitable materials. The choice of material will probably lie between the use of a paint, such as chlorinated rubber paint, or a PVC/aluminium foil-backed paper. It is important that the material is carried

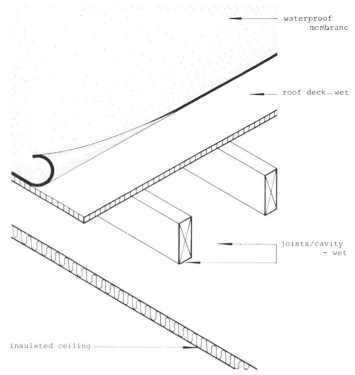

6 *Condition 3: Cold deck roof (1)*

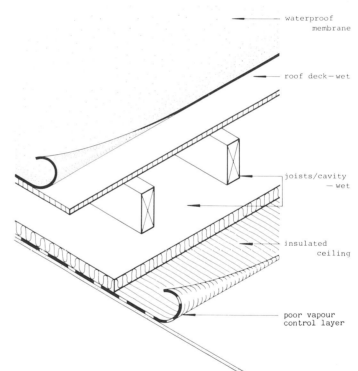

7 *Condition 4: Cold deck roof (2) vapour control below insulation*

down the external walls for a distance of at least 200 mm from the wall/ceiling junction.

Condition 4: Cold deck roof (2)

Cause
1 The cavity is inadequately ventilated.
2 The vapour control layer is poor (i.e. it has insufficient vapour resistance or is perforated, at light drops or services outlets especially).

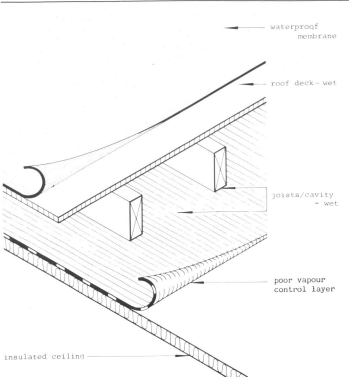

waterproof membrane

roof deck - wet

joists/cavity - wet

poor vapour control layer

insulated ceiling

8 *Condition 4: Cold deck roof (2) vapour control above insulation*

Remedial work
1 *At best*:
Increase the cavity ventilation so that the ventilation openings are on opposite sides of the roof and are equivalent to *at least* 0.4 per cent of the plan area – at least 0.6 per cent of the plan area for a low building or a building in a sheltered location. Where there are cavity barriers in the roof, as required by the Building Regulations, cowl ventilators will have to be installed to achieve adequate ventilation of the isolated zones.
2 *At worst*:
In addition to increasing the cavity ventilation, remove the existing ceiling and vapour control layer and reinstate the ceiling with new thermal insulation and a vapour control layer on the warm side. If Gyproc vapour check wallboard is used, it is advisable to seal all the joints between the boards with a vapour-resistant tape such as Densotape manufactured by Winn & Coales Limited or with a gun-applied sealant compatible with the polythene film backing. Depending on the vapour resistance required, consideration could be given to the use of Styroliner (Sheffield Insulation Limited). Styroliner is a plasterboard/extruded polystyrene composite. The extruded polystyrene has considerable vapour resistance. If used, it is important that the joints between boards are sealed with a vapour resistant tape (e.g. Densotape) or filled with a gun-applied sealant.

References for correct detailing
1 General:
(a) BRE Digest 312, *Flat roof design: the technical options,* gives comprehensive guidance on the principles of warm and cold deck roofs. No specific details are given.
(b) NHBC Practice Note 13, 1980, *Construction of flat roofs and pitched roofs with a fully supported continuous weatherproofing membrane* gives very clear guidance on

what and what not to do. Bold, clear diagrams are used throughout.
(c) BRE Information Sheet IS 6/78, *Considerations in the design of timber flat roofs.* A good general discussion of the options, emphasising the risk of condensation, with schematic details.
(d) GLC, *Development and Materials Bulletin* **116**, June 1978, item 6, 'Condensation prevention in high moisture production areas'. This is a case study in which the need for adequate removal of saturated air by ventilation – by mechanical means if needs be – is reiterated. *Condensation, AJ Construction Risks and Remedies* series 9.4.86, pp. 49–58 and 16.5.86, pp. 69–81, discusses the basic mechanics of condensation, the causes of the most important problems and how they can be avoided or resolved.
2 For warm roofs:
(a) BS 5250: 1989: *Code of Practice for Conrol of condensation in buildings,* clause 9.4.3.2, page 18 defines the 'warm deck'; design guidance is given on particular roof constructions having a timber deck in 9.4.8.2 and 9.4.8.2, page 26: Note: The revised code is completely different in format, coverage and extent than the 1975 code that it now supersedes.
(b) PSA, *Technical guide to flat roofing,* 1987, 2.10.3, page 31 (definition) and 2.10.13, page 37 and 2.10.13–16 pages 37–42, (surface and interstitial condensation in warm and inverted roofs).
(c) Tarmac's *Flat roofing, a guide to good practice,* 1.2 'Thermal design' and 1.3 'Vapour design guide', is the first to introduce in book form the concept of moisture gain analysis by which it may not be necessary always to have a vapour control layer in a warm deck flat roof design. No methods of prediction nor properties are given but it has useful tables and a summary covering a wide range of constructions and internal temperature conditions.
(d) BS 6229: 1982: *Code of practice for flat roofs with continuously-supported coverings,* is the first Code of Practice to introduce the concept of moisture gain analysis (see also reference (c) above). Appendices give method of prediction, vapour resistance properties (described differently from existing references) and examples.
(e) BRE Information paper 2/89, *Thermal performance of lightweight inverted warm deck roofs,* January 1989. Instructive and demonstrates the importance of thermal capacity in inverted roofs
(f) BRE Diges 336, *Swimming pool roofs: minimising the risk of condensation using warm-deck roofing,* September 1988, underlines the value of warm deck designs, for high risk condensation building types/areas in particular.
3 For cold roofs:
(a) BS 5250: 1989: *Code of Practice for Control of condensation in buildings,* clause 9.4.3.1, page 17 (definition) and 9.4.8.1, page 25 (design guidance)
(b) PSA *Technical guide to flat roofing,* 1987, 2.10.4, page 32 (definition) and 2.10.17, page 43 (interstitial condensation).
(c) BRE Digest 218: *Cavity barriers and ventilation in flat and low pitched roofs,* sets out the requirements of the Building Regulations concerning barriers. Fig. 1 on page 3 gives details for achieving ventilation using cowl ventilators. The free area of the ventilation openings should be equivalent to *at least* 0.4 per cent of the plan area of the roof – but see (f) below for suggested increase to 0.6 per cent.

(d) BRE Information Paper IP 35/79 *Moisture in a timber-based flat roof of cold deck construction,* describes a simulation experiment which further demonstrated the need for ventilation of the roof void in a 'Cold roof' – but see (f) below for latest advice.

(e) Tarmac's *Flat roofing, a guide to good practice.*

(f) BS 6229: 1982: *Code of practice for flat roofs with continuously-supported coverings* recommends minimum ventilation aperture of 0.4 per cent of the roof plan area – but see (f) below for qualifications in certain circumstances.

(g) BRE Information Paper, IP 13/87, *Ventilating cold deck flat roofs,* explains why there may be a need to increase current minimum aperture of 0.4 per cent of the plan roof area (BS 6229) to 0.6 per cent for complex plan shapes or for simple plan shapes of low buildings or buildings located in a sheltered location.

(h) BRE DAS 59, *Felted cold deck flat roofs: remedying condensation by converting to warm deck,* is a succinct text, good on precautions and with helpful three-dimensional illustrations.

Study 9
Flat roofs: parapets and abutments

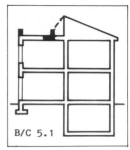

B/C 5.1

Manifestation 1

1 *At best*:
Damp staining of and persistent efflorescence on the brickwork below the coping of the parapet.
2 *At worst*:
Frost splitting of the brickwork and/or sulphate attack of the mortar joints.

Cause

The parapet is excessively wet because:

1 The dpc under the coping is absent or ineffective (**1, 2**). The worst condition occurs if there is no dpc. The absence of a dpc across the cavity allows direct penetration of rainwater into the cavity, particularly between the joints and the coping; a sagging dpc allows penetration of rainwater into the cavity at laps or splits in the dpc material.
2 There is inadequate protection of the coping and/or there are no drips or the drips are faulty.
3 Weepholes above the dpc are absent and this has led to water collecting and saturating the brickwork rather than draining away.

Cross references

Study 22: Rain penetration under 'Dpcs/parapets', 6.05, page 123.
Study 24: Movement, page 134 ff.

Remedial work

1 *At best*:
Remove and reinstate the coping with a proper dpc beneath it. The dpc should project about 10 mm beyond the face of the brick and should be fully supported across the cavity by slates or asbestos cement tiles Introduce movement joints, 1.5 mm from corners and then at intervals of 4 m minimum.
2 *At worst*:
Take down the parapet and rebuild it, incorporating a proper dpc (and, if necessary, dp tray) with all laps fully bonded (and if a new dp tray is also installed ensure that it laps over the top of the skirting of the waterproof membrane). Introduce movement joints as for 1 above.

References for correct detailing

1 Handisyde, C. C., *Everyday details*, 21, 'Parapets in masonry construction', page 99 ff.
2 Duell and Lawson, *Damp-proof course detailing*, Fig. 21a, page 14.
3 BRE DAS 106, *Cavity parapets – avoiding rain penetration*, and DAS 107, *Cavity parapets – installation of dpcs, trays and flashings*, both have succinct text and good diagrams, some in three dimensions.
4 PSA Feedback, *Single storey extension – cavity wall defects*, Building File 2, page 43.
5 BS 5642: 1983: Part 2, *Sills and copings*, also gives

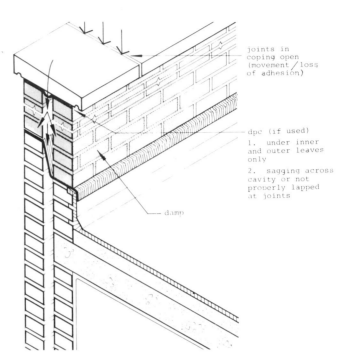

joints in coping open (movement / loss of adhesion)

dpc (if used)
1. under inner and outer leaves only
2. sagging across cavity or not properly lapped at joints

damp

1 *Rainwater penetrating ineffective dpc immediately below parapet coping.*

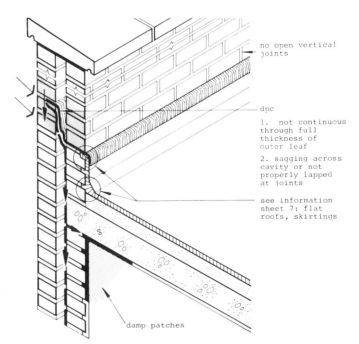

no open vertical joints

dpc
1. not continuous through full thickness of outer leaf
2. sagging across cavity or not properly lapped at joints

see information sheet 7: flat roofs, skirtings

damp patches

2 *Rainwater bridging cavity in parapet above the waterproof membrane, due to ineffective dpc.*

48

recommendations for design and installation of coping systems and supersedes BS 3798: 1964.

6 BS 5628: Part 3: 1985: *British Standard Code of Practice for use of masonry,* Part 3 Materials and components, design and workmanship (formerly CP 121: Part 1). For rainwater exclusion, Fig. 12, page 51 related to clauses 21.4, 21.5.7 and 21.7; for frost action, clause 22.3; and for sulphate attack clause 22.4.

7 BRE Digest 89, *Sulphate attack on brickwork* (minor revisions, 1971).

8 BRE Digest 157, *Calcium silicate (sandlime, flintlime) brickwork.*

9 BRE Digest 200, *Repairing brickwork.*

10 Nash, W. G., *Brickwork repair and restoration,* 6 'Repairing and restoring brick features', is a good practical treatment (as is the rest of the book) on most traditional parapets and could be especially useful in rehabilitation/refurbishment and is included here accordingly.

11 PSA *Technical guide to flat roofing,* 1987, 3.6.5, Figure 3.22, page 36 and Figures 3.24 and 3.25, page 38 and 3.10.5, Figure 3.129, page 125, Figure 3.130, page 126 and Figure 3.131, page 127. (Similar details but included under asphalt and built-up felt roofs respectively).

Manifestation 2
Dampness internally at or near the junction of the wall and ceiling.

Note: Blistering of the skirting of the waterproof membrane prior to dampness occurring internally is usually indicative of damp brickwork behind the membrane.

Cause 1
If associated with periods of heavy rain, rainwater may be bypassing the ineffective dpc in the following ways:

1 In the outer leaf the dpc does not extend to the face of the wall so that water can bridge it and then trickle down the underside of the sloping dpc to the inner leaf.

2 Faulty laps or other openings in the sloping dpc provide a path for water from the outer leaf directly to the inner leaf. This problem is aggravated by the absence of weepholes in the exposed portion of the inner leaf, since water can dam up and then find its way through the sloping dpc. The exposed nature of parapets makes them particularly vulnerable to shortcomings in design or workmanship.

3 The roof membrane skirting is bridged by accumulated debris.

Cross reference
Study 22: Rain penetration, 'Dpcs/Parapets', 6.05, page 123.

Remedial work
If trees are close by ensure that debris is removed regularly, particularly in the autumn.

1 *At best*:
(a) If the dpc is sound but there are no weepholes, form weepholes at every fourth perpend, making sure that the dpc is not damaged in the process.
(b) If faults in the dpc are localised, remove a short length of the parapet in the vicinity of the fault in the dpc, repair the dpc and then make good to the parapet. Form weepholes if necessary (as in (a) above). It may be advisable to introduce movement

joints in the joints of the coping, 1.5 m from corners and then at intervals of 4.0 m minimum.

(c) Alternatively, clad the inner face of the parapet so that there is continuous damp-proofing from the skirting of the roof to the front of the coping. Faulty dpcs can be left undisturbed, since they will now be redundant.

Note: Use of a Borescope enables a survey of the condition of the dpc to be made without dismantling the parapet.

2 *At worst*:
Take down the parapet and rebuild it, incorporating a proper dp tray with all laps fully bonded and ensuring that the dp tray laps over the top of the skirting of the waterproof membrane and introduce movement joints, 1.5 m from corners and then at intervals of 4 m minimum.

References for correct detailing
1 Handisyde, C. C., *Everyday details,* 21, page 102.
2 Duel and Lawson, *Damp-proof course detailing,* Fig. 21b, page 14.
3 Eldridge, H. J., *Common defects in buildings,* Figs. 9 and 10, pages 53–55.
4 DOE, *Construction* **29**, March 1979, page 1, 'Cavity trays in cavity walls'.
5 BRE DAS 106, *Cavity parapets – avoiding rain penetration,* and DAS 107, *Cavity parapets – installation of dpcs, trays and flashings,* both have succinct text and good diagrams, some in three dimensions.
6 PSA Feedback, *Single storey extension – cavity wall defects,* Building File 2, page 43.
7 BS 5642: 1983: Part 2: *Sills and copings,* also gives recommendations for design and installation of coping systems and supersedes BS 3798: 1964.
8 BS 5628: Part 3: 1985: *British Standard Code of Practice for use of masonry,* Part 3 Materials and components, design and workmanship (formerly CP 121: Part 1). Fig. 12, page 51 related to clauses 21.4, 21.5.7 and 21.7.
9 PSA *Technical guide to flat roofing,* 1987, 3.6.5, Figure 3.22, page 36 and Figures 3.24 and 3.25, page 38 and 3.10.5, Figure 3.129, page 125, Figure 3.130, page 126 and Figure 3.131, page 127. (Similar details but under asphalt and built-up felt roofs).

Note: All references, except 9, show the cavity tray falling towards the roof. The argument for this is that the weepholes do not drain onto the face of the wall which would cause unsightly staining. The argument against is the risk of water travelling via the underside of the dpc to the interior of the building (for the reasons given above); this risk is increased where there is cavity insulation. It has also been argued that condensation forming on the underside of the dpc would drip down onto the inner leaf. However, the PSA reports no failures of either kind, but appears to prefer having the dp tray falling to the outside:- see 9 above.

Cause 2
If not associated with rain, surface condensation due to cold bridging.

Note: Mould growth at the wall/ceiling junction may be the only manifestation of dampness.

Cross reference
Study 20: Condensation under heading 'Surface condensation: cold bridges', **4.02**, page 108.

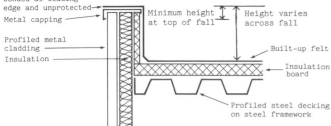

3 *Condition leading to wind stripping of felt to parapet of profiled metal wall cladding due to inadequate fixing/ protection of the felt – see* **4a, b**.

4a, b *Poor bonding of felt taken over capping to profiled metal wall cladding,* **a,** *with the felt and fibreboard to the perimeter upstand pulled away illustrating minimal adhesive at the top,* **b**

Remedial work
1 *At best*:
Increase ventilation or increase heating, and/or clean off the mould growth with an 'anti mould wash' (e.g. one part bleach to four parts water), and apply Silexine's Anticon to the ceiling and Fungichek to the walls, all in accordance with the maker's recommendations.
2 *At worst*: Increase the thermal insulation of the wall/ ceiling junction by applying an adequate thickness of thermal insulation for a distance of at least 450 mm along the ceiling and 300 mm down the wall. The use of extruded polystyrene, such as Styrofoam, has the advantage over other insulants in that the material itself has the properties of a vapour control layer. If other insulants are proposed which are vapour permeable, such as expanded polystyrene, a vapour control layer on the warm side will be required. Composite panels of extruded polystyrene and plasterboard (Styroliner from Sheffield Insulation) and of expanded polystyrene, polythene film and plasterboard (Gyproc thermal board from British Gypsum) are now available. Whatever board is used, it is absolutely essential that the boards are uniformly bonded to the existing ceiling or wall so as to reduce the risk of insterstitial condensation forming at the wall or ceiling and thermal insulation interface. All joints between boards should be sealed either with a vapour-resistant tape such as Densotape by Winn and Coales Limited or with a gun-filled sealant compatible with the underlying materials. The thickness of the insulant will probably not be less than 15 mm. The minimum amount of thermal insulation required can be calculated by following the methods described in the references listed at the end of Study 20, Condensation, 5, page 112ff. Increased ventilation and/or higher levels of sustained heating may also be necessary.

Manifestation 3
Stripping of built-up felt from parapet to profiled steel external cladding and sometimes with stripping of built-up felt and insulation from adjacent flat roof as well.

Cause
Inadequate fixing of felt at leading edge of top of parapet and/or lack of protective capping.

Cross reference
Study 24: Movement, particularly note on wind effects, page 135.

Remedial work
Strip off felt to top and sides of parapet and for a short distance (say, 300 mm) along flat roof. Reinstate with new mineral surfaced or metal foil-faced felt, bonded to existing along flat roof and mechanically fixed to top of parapet. Cover top of parapet with a pressed metal capping fixed to the wall cladding. Ensure that butt joints along the capping are properly butt strapped and sealed.

References for correct detailing
1 Tarmac's *Flat roofing, a guide to good practice,* 'Skirting to cladding parapet', page 166.
2 CP 144: Part 3: 1970: *Built-up bitumen felt*, Fig. 9, page 34.
3 PSA, *Technical guide to flat roofing,* 1987, 3.10.6, Figure 3.140, page 131 (illustrates a verge that is the same in principle as a parapet).

Note: This reference is included for convenience. If used, the details illustrated should be adapted for a metal deck.

5 *Outward movement of brick parapet at corner; and* **6,** *detachment and cracking of felt skirting to timber roof. Both due to excessive movement and insufficient movement joints.*

Manifestation 4

Vertical or diagonal cracking (of units and/or mortar joints) with or without dampness internally at or near the junction of the wall and ceiling.

Cause
Thermal and/or moisture movements. The parapet and/or roof length is excessive for the degree of exposure to heat and/or rain and for the properties of the units and/or the mortar.

Note: Frost action and/or sulphate attack may be responsible for or may have contributed to the total movement.

Cross references
Study 24: Movements, particularly the diagnostic checklist for thermal movement, 7, page 138, and moisture movement, 8, page 140.

Remedial work
1 *At best*:
None is required if the parapet is safe and there is no water penetration.

2 *At worst*:
(a) If the parapet is unsafe: rebuild incorporating correct movement joints.
(b) If the parapet is safe but allows water penetration, dpcs must be checked and replaced if damaged (see Manifestation 2, Remedial work for cause 1, earlier). Rebuilding may be necessary. If the parapet is to be rebuilt (and in some cases even if it is not being rebuilt), it is wise to detach the skirting of the waterproof membrane from the parapet, and sealing the top of the parapet with an apron flashing.

References for correct detailing
1 For movements in parapets:

(a) BS 5628: Part 3: 1985: *British Standard Code of Practice for use of masonry*, Part 3, *Materials and components, design and workmanship* (formerly CP 121: Part 1). Thermal movements of masonry units in Table 19, page 87 and Moisture movements of masonry units in Table 20, page 88. General rules for spacing of movement joints: Fired-clay masonry, 12 m and should never exceed 15 m (clause 20.3.2.2); Calcium silicate masonry between 7.5 and 9 m (clause 20.3.2.3); and Concrete masonry, 6 m (clause 20.3.2.4). Fig. 10 on type and location of movement joints.
(b) Eldridge, H. J., *Common defects in buildings*, pages 152–155, emphasises the cumulative effect of the causes.

7 *Detachment of skirting due to movement; note crack along top edge of skirting.*

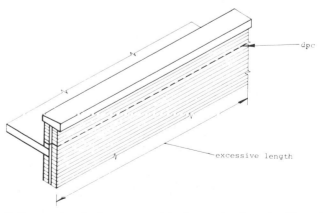

8 *Excessive length of parapet, resulting in excessive thermal and/or moisture movement.*

(c)   BRE Digest 89, *Sulphate attack on brickwork,* describes the basic chemistry, the symptoms and the precautions to be taken in new work. Basically, these are to use sulphate-resistant mortars in exposed situations and to protect exposed brickwork by drips, copings and flashings.

2   For rebuilding parapets:

(a)   Handisyde, C. C., *Everyday details,* 21, page 102.

(b)   Duell and Lawson, *Damp-proof course detailing,* Fig. 21b, page 14.

(c)   BS 5628: Part 3: 1985: *British Standard Code of Practice for use of masonry,* Part 3: *Materials and components, design and workmanship* (formerly CP 121: Part 1), Fig. 12, page 51.

(d)   Eldridge, H. J., *Common defects in buildings,* Figs. 9 and 10, pages 53–55.

(e)   DOE, *Construction,* **29,** March 1979, page 1, 'Cavity trays in cavity walls'.

(f)   PSA *Technical guide to flat roofing,* 1987, 3.6.5, Figure 3.22, page 36 and Figures 3.24 and 3.25, page 38 and 3.10.5, Figure 3.129, page 125, Figure 3.130, page 126 and Figure 3.131, page 127. (Similar details but included under asphalt and built-up felt roofs respectively). Illustrates basic construction and requirements for movement joints.

Note: All these references, except (f), show the cavity tray falling towards the roof. The argument for this is that the weepholes do not drain on to the face of the wall which would cause unsightly staining. The argument against is the risk of water travelling via the underside of the dpc to the interior of the building (for the reasons given above); this risk is increased where there is cavity insulation. It has also been argued that condensation forming on the underside of the dpc would drip down on to the inner leaf. However, the PSA reports no failures of either kind, but appears to prefer having the dp tray falling to the outside – see (f) above.

Manifestation 5

Dampness internally at or near the abutment of the flat and pitched roof.

Cause

Vortex action by the wind (0) that is particularly severe when the sloping abutment is facing in the direction of the prevailing wind. The effect of the vortex is:

(a)   to drive rainwater up laps further than would occur by normal head of water criteria, and

(b)   to pump rainwater into and over laps at joints where the waterproofing membrane is of a lightweight metal cladding (such as aluminium, stainless steel or zinc).

Cross references

Vortex action of the wind at the junction of a flat roof and a sloping abutment has not been the subject of much research, and, where studied (usually specifically in connection with particular cases of unusual roof geometry), the results have not been publicised.

Remedial work

1   *At best*: Where the case is one of built-up felt or asphalt waterproofing having been taken up behind tile or slate cladding, making the lap of the flat roof waterproofing longer will usually suffice.

2   *At worst*: Where the case is one of lightweight cladding being the waterproofing membrane on both flat and sloping roofs, the remedy will almost certainly involve a change in the design at the junction of the two roofs and most likely the use of a jointless waterproofing membrane at the junction. There may be a need for a wind tunnel test to determine the severity of the problem and hence the nature of the remedial work in detail.

References for correct detailing

Consult specialists in wind tunnel tests.

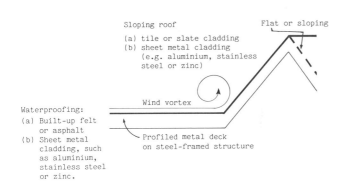

9a *Diagrammatic section of a flat roof/sloping roof abutment that can give rise to rainwater penetration at or near the junction because of vortex during wind.* **b** *Photographic example where vortex action was a contributory factor.*

Study 10
Flat roofs: verges

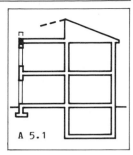
Manifestation 1
Cracking and splitting of the waterproof membrane.

Cause
Differential movement – moisture or thermal – between the substrate and the membrane or between the membrane and the edge condition (usually the trim).

Note: The precise cause is often difficult to determine. Failure seems to take place more commonly, though not exclusively, at or near both internal and external corners, and movement of the wall itself may be strongly involved.

Cross reference
Study 24: Movement, particularly diagnostic checklist for thermal movement, 7 page 138.

Remedial work
Entirely dependent on positive identification of the cause (and such identification is often difficult).

1 *At best*:
   (reforming the edge condition):
   (a) If there are cracks and splits in the roof covering at joints in aluminium roof trim or the trim has become dislodged or disturbed: refix the roof trim in 1300 mm maximum lengths, fastened at 300 mm maximum centres with 3 mm expansion gap between lengths of the trim. Insert an internal jointing sleeve below the joints, the sleeves locked on one side only to permit movement. Make good to the roofing that has been damaged or disturbed.
   (b) If there are cracks or splits in the roof covering along the line of joint at change in materials of roof deck and the eaves construction: strip the affected roofing at the eaves, insert a minor movement joint consisting of a strip of high-performance felt over the joint in the base, bonded only at the edges and make good to the roofing.
   (c) If there are cracks or splits in asphalt roofing along the line of the back edge of a metal flashing or at welted or lapped joints, or there is separation of the asphalt from the flashing: strip and refix the flashing in a recess to maintain the full thickness of asphalt

over the welted back edge and joints. The flashing should be securely fixed to the base. Make good to the damaged asphalt.

2 *At worst*:
   A substantial part of the roof may require to be reinstated with appropriate movement joints incorporated, as described in Study 6, Manifestation 3, 'Remedial work', page 33.

References for correct detailing
Note: Existing Codes of Practice (CP 144) do not give guidance on the use of some of the newer trims. Be sure to consult with the roofing contractor and/or the manufacturer of the trim first.

3 *Cracking of asphalt on precast concrete roof deck, due to differential movement in structure underlying finish.*

4 *Split in felt at change in direction of roof trim.*

5 *Crack in felt at joint in roof trim, due to thermal movement of parapet wall. (All photos Building Research Establishment, Crown Copyright)*

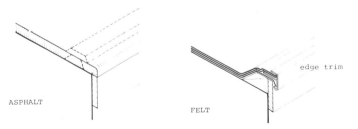

ASPHALT

FELT

edge trim

1, 2 *Two forms of edge treatment which may be encountered; one an asphalt roof finish, the other one roofing felt.*

1 MACEF, *Roofing handbook*, figures on pages 27–29.
2 FRCAB, *Roofing handbook*, June 1988, Information shet 1, 'Verge detail' trim with built-up felt.
3 Handisyde, C. C., *Everday details* 20, particularly Figs. 45–49 on page 98. This deals only with timber substructure. The need to allow for thermal expansion of metal or plastic trims by providing flashing strips underneath the joints is emphasised.
4 Tarmac's *Flat roofing, a guide to good practice*: for built-up felt, Metal trims, page 154; for asphalt, Eaves and verges (principles), pages 176 and 177, and (details) page 185.
5 PSA, *Technical guide to flat roofing*, 1987, 3.6.6, Figures 3.29 to 3.38, pages 40 to 45 (asphalt) and 3.10.6, Figure 3.134 to 3.141, pages 128 to 132. Comprehensive coverage including welted drips and trims for a different deck and inverted roof.

Manifestation 2

Detachment and/or stripping of built-up felt, sometimes with the adjacent underlying insulation boards. Compare with Parapets, Study 9, Manifestation 3, page 50.

Cause
Inadequate fixing of the edge of the felt.

Cross reference
Study 24: Movement, particularly note on wind effects, page 135.

Remedial work
Reinstate verge with adequate fixing of edge. Built-up felt and/or insulation boards adjacent to the verge may also require reinstatement.

References for correct detailing
1 CP 144: Part 3: 1970: *Built-up bitumen felt*, clause 3.3, page 13, for advice on wind suction and Figs. 7 (welted drip) and 8 and 9 (metal trims), page 34.
2 Tarmac's *Flat roofing, a guide to good practice*, Eaves and verges, pages 153 and 155 (for welted drips and metal trims). Includes advice on wind suction to be expected at verges.
3 PSA, *Technical guide to flat roofing*, 1987, 3.6.6, Figures 3.29 to 3.38, pages 40 to 45 (asphalt) and 3.10.6, Figures 3.134 to 3.141, pages 128 to 132. Comprehensive coverage including welted drips and trims for a different deck and inverted roof.

Study 11
Flat roofs: skirtings

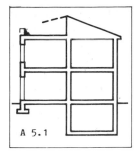

Manifestation 1

1 Internally: dampness at or near wall/ceiling junction.
2 Externally: pointing above the asphalt has fallen out and/or the asphalt has dislodged slightly.

Cause

Rainwater has found its way behind the asphalt. Detachment and/or dislodgement of the pointing and/or asphalt has occurred because the horizontal chase was inadequately formed and/or sized (i.e. it is less than 25 mm × 25 mm with the bottom splayed) for proper asphalt tucking and mortar pointing above (**1, 2**).

Cross references

Study 22: Rain penetration under heading 'Asphalt skirtings', 6.03, page 123.
Study 25: 'Asphalt skirtings to flat roofs', 12, page 154.

Remedial work

Remove the asphalt, 50–60 mm below the existing chase. Cut a new chase correctly sized and shaped and then relay the asphalt with a weathered top and space for pointing.

1 *Poor tuck-in and pointing of asphalt skirting.*

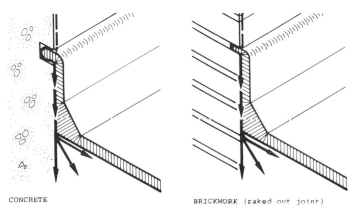

CONCRETE   BRICKWORK (raked out joint)
any type of roof·deck in both cases

2 *Rainwater finding its way behind the asphalt skirting via inadequate chase detail, leading to dampness at wall/ceiling junction beneath.*

Point with cement:sand (1:3) between the top of the asphalt and the underside of the chase.

References for correct detailing

1 CP 144: Part 4: 1970: *Roof coverings: mastic asphalt*, Fig. 1, page 28 shows the detail for brickwork (but without the splay, as does reference 2 below – and it is wiser to incorporate the splay) and Fig. 2, page 29, the correct detail for concrete where the deck is also concrete and therefore stable. Where there is likely to be differential movement between the upstand and the deck a detail incorporating a free-standing kerb should be used as in Fig. 4, page 31.
2 MACEF, *Roofing handbook*, the details on pages 15, 16 and 17 show the correct details when the deck is stable. The treatment using free-stand kerbs is shown on pages 12, 13 and 18.
3 PSA, *Technical guide to flat roofing*, 1987, 3.6.1, Figure 3.8, page 29 (basic requirements with 'vapour barrier': warm deck), 3.6.3, Figures 3.13 to 3.19, pages 32 to 35 (includes associated dp trays, uninsulated and insulated walls for warm deck and inverted roofs but for concrete decks only, unstable decks not included but see MACEF reference above. For unstable decks note in particular minimum skirting height of 200 mm).
4 Tarmac's *Flat roofing, a guide to good practice*, 5.3 'Mastic asphalt detail design', in particular pages 170–172 is the best of the references. The principles are explained and details clearly drawn.

Manifestation 2

Externally: Slumping and/or cracking of skirtings or deformation of fillet or all three.

Note: If the skirting has cracked or has pulled out of the chase there could also be dampness internally at or near the wall/ceiling junction as in Manifestation 1 above.

Cause

1 Inadequate thickness of the vertical asphalt (i.e. less than 13 mm or 19 mm when the skirting is greater than 300 mm high).
2 The vertical substrate does not provide an adequate key for the asphalt, either by being too smooth or by being damp at the time the asphalt was applied.
3 The asphalt became too soft owing to the lack of solar reflective treatment.
4 The angle fillet was too small (i.e. less than the minimum of 50 mm).
5 The skirting asphalt was not adequately supported at its base, in particular where a board foam or compressible board has been used for the insulation (**6**).

Cross references

Study 22: Rain penetration, under heading 'Asphalt skirtings', 6.03, page 123.
Study 25: 'Asphalt skirtings to flat roofs', 12, page 154.

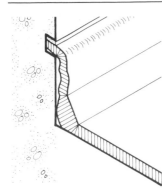
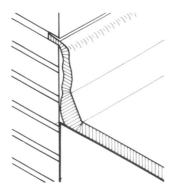

CONCRETE   BRICKWORK
all types of roof deck and walls

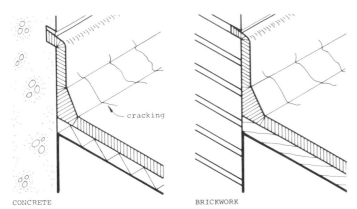

CONCRETE   BRICKWORK
woodwool/timber roof deck

3 *Slumping of skirting; if excessive, this can lead to skirting pulling out of chase and consequent damp penetration.*

4, 5 *Slumping of skirting. (Photos Building Research Establishment and GLC.)*

Remedial work

Remove the asphalt for the full height vertically and horizontally for about 50 mm from the angle fillet. Provide an adequate key to the vertical substrate. For concrete the surface should be roughened; for brickwork the joints should be brushed off rather than deeply raked out. In addition, cement:sand gauged with PVAC; or bitumen rubber emulsion can be applied to the vertical surface. The asphalt should be relaid on isolating felt where it is horizontal so that it is level with the existing asphalt; 13 mm minimum thickness of asphalt should be laid vertically with a 50 mm minimum angle fillet. Where the deck is likely to move in relation to the wall, use a free-standing kerb detail, together with an apron flashing. (See Manifestation 3 below.)

References for correct detailing

1 CP 144: Part 4: 1970: *Roof coverings: mastic asphalt*, Fig. 1, page 28, shows the detail for brickwork (but without the splay, as does reference 2 below – and it is wiser to incorporate the splay) and Fig. 2, page 29, the correct detail for concrete where the deck is also concrete and therefore stable. Where there is likely to be differential movement between the upstand and the deck, a detail incorporating a free-standing kerb should be used as in Fig. 4, page 31.

  The code does not give advice where the deck or substrate for the asphalt is unstable – see reference 3 below for the advice needed.

2 MACEF, *Roofing handbook*; the details on pages 15, 16 and 17 show the correct details when the deck is stable; and pages 17 (lower detail) and 18 (lower detail) the deck or substrate for when the asphalt is unstable. The treatment using free-standing kerbs is shown on pages 12, 13 and 18.

3 PSA, *Technical guide to flat roofing*, 1987, 3.6.1, Figure 3.8, page 29 (basic requirements with 'vapour barrier': warm deck), 3.6.3, Figures 3.13 to 3.19, pages 32 to 35 (includes associated dp trays, uninsulated and insulated walls for warm deck and inverted roofs but for concrete decks only. For unstable decks note in particular minimum skirting height of 200 mm).

4 Tarmac's *Flat roofing, a guide to good practice*, 'Skirting support', pages 173 and 174 and for details, pages 178 and 179 (top details).

Manifestation 3

Externally: Extensive cracking in the region of the angle fillet.

Cause

Differential movement between the roof deck and the upstand or parapet, sometimes aggravated by the fillet being too small (**8**).

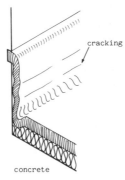

6 *Deformation of fillet due to lack of support of skirting from the underlying insulation leading to slumping with or without cracking – see* **7**.

7 *Deformation of fillet into a rounded shape because the asphalt was not fully supported by the insulation below it. Cracking of skirting from the combined effects of the inadequate bonding of the asphalt to the brickwork – the joints in the latter were not raked.*

9 *Built-up bitumen felt badly formed at corner; pipe flashing without cover flashing.*

Cross references
Study 24: Movements, under heading 'Movement joints', 8.05, page 144 and Fig. 25.

Remedial work
Remove the asphalt for the full height vertically and horizontally for about 150–200 mm. Form a pre-screeded woodwool or timber kerb with a 13 mm minimum movement gap between the back of the kerb and face of the upstand or parapet. Tack expanded metal lath on the face of the kerb and for 100 mm along the horizontal surface over new isolating membrane. Renew asphalt. Install a non-ferrous metal apron flashing with cleats and dress over the top of the asphalt to the kerb.

References for correct detailing
1 CP 144: Part 4: 1970: *Roof coverings: mastic asphalt*, Fig. 4, page 31.

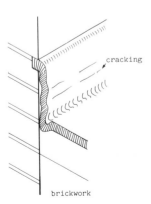

8 *Cracking due to differential movement, possibly letting through rainwater to ceiling/wall junction below.*

2 MACEF, *Roofing handbook*. The figure on page 12 shows the detail with a timber kerb. With a concrete kerb, no expanded metal is used and the cleat as well as flashings chased into the upstand. However, expanded metal on isolating felt should be used on lightweight or very smooth concrete.
3 NHBC Practice Note 13, *Construction of flat roofs*. Detail under clause 4(g) on page 10.
4 Tarmac's *Flat roofing, a guide to good practice*, for free-standing kerbs with metal decking only, pages 184 and 185.

Manifestation 4
Internally: dampness associated with projections through the roof such as rooflights, ducts and pipes or rooftop additions laid on the waterproof membrane.

Cause
1 The skirting height inadequate (i.e. is less than 150 mm) and/or cover flashing incorrectly fixed or missing.
2 Waterproofing badly formed at corners (with built-up felt or single membrane roofings).
3 Inadequate or lack or cover flashing (to ducts and pipes).
4 Siting of ducts and/or pipes precludes proper waterproofing.
5 Building services (pipes or ducts) taken unsupported across the membrane; tanks, housings, condensers, ventilation and similar units stood directly on the membrane; and aerials and signs stood directly on the membrane.

Cross reference
Study 22: Rainwater penetration under heading 'Projections through roofs', 6.06 page 124.

10, 11 *Site of duct and pipes such as to preclude proper waterproofing (asphalt in this case),* **10,** *space between rooflights too narrow to have felt laid properly,* **11** *– ponding within space too!*

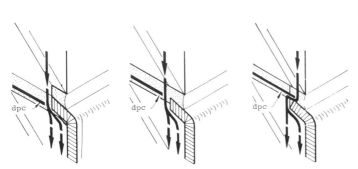

12 *Three conditions in chase where dpc meets roof upstand tuck-in, allowing water penetration.*
13 *A built example of the condition illustrated.*

Remedial work
According to cause:

1 Raise rooflight, relay asphalt with proper cover flashing.
2 Strip membrane and relay with proper folds and laps.
3 Fix proper cover flashing (it may be necessary to strip and reform the membrane).
4 Resite the projection(s) so that there is adequate space around each projection to enable the roofer to lay the membrane and provide a cover flashing. It may be possible to group pipes into a box and then to waterproof the box. This would still involve resiting the pipes. There is little else that can be done if continual maintenance is to be avoided – an important aim of remedial work.
5 Provide support for the pipes or ducts that distributes the load over the membrane, if necessary providing a separate pad or frame (for aerials or signs, for example) or a separate structural kerb (for tanks, housings, condensers, ventilation units, for example) that can be properly waterproofed.

References for correct detailing
1 Tarmac's *Flat roofing, a guide to good practice:* for built-up felt, skirtings to cold pipes, hot pipes and flues, page 152

(details on pages 162, 165 and 168); skirtings to metal rooflight, pages 161 and 164; rooftop additions, pages 158 and 159 (principles only). For asphalt: skirtings to all projections, pages 174 and 175 (details for metal decks (kerbs and pipes)), page 186.
This is the clearest, most comprehensive reference.
2 PSA, *Technical guide to flat roofing*, 1987. For asphalt: rooflights and trapdoors, 3.6.8, Figures 3.41 and 3.42, pages 46–47; flues, pipes and services, 3.6.10, Figures 3.52–3.58, pages 52–58; for railings, safety barriers, etc., 3.6.11, Figures 3.59–3.66, pages 58–62al with an inverted roof. For felt: rooflights and trapdoors, 3.10.8, Figures 3.144–3.146, pages 133 and 134; flues, pipes and services, 3.10.10, Figures 3.151–3.156, pages 138–143, Railings, safety barriers and lightning conductors, 3.10.11, Figures 3.157–3.163, pages 143–147.
PSA, *Flat roofs, Technical Guide* (2nd Edition). Remedial work: skirtings generally, clauses 311.9, 10, 11 and 12, page 63; fixings and bases, clause 314, pages 67 and 68; rooflights and trapdoors, clause 317, pages 70 and 71; pipes and vents, clause 318, page 73.
3 CP 144: Part 3: 1970: *Built-up bitumen felt:* for skirtings generally, Figs. 1–5, pages 32 and 33; for pipe and duct projections, Figs. 10–12, page 35.

4 CP 144: Part 4: 1980: *Mastic asphalt*, for wood and metal sills with concrete kerbs, Figs. 11 and 12, pages 36 and 37; pipes and standards, Figs. 18 and 19, pages 44 and 45.

5 MACEF, *Roofing handbook*: for kerbs to rooflights, pages 21 and 22. No details of other projections but page 34 has clear detail for asphalt plinths with a metal flashing for cradle tracks or equipment.

Manifestation 5
Internally: dampness at or near the wall/ceiling junction.

Cause
The dpc was not taken over the top of the splayed asphalt and/or not extended beyond the face of the vertical asphalt, thus allowing water to penetrate behind the asphalt **12**.

Cross reference
Study 22: Rain penetration under heading 'Dpcs/parapets', 6.05, page 123.

Remedial work
A modified form of apron flashing (a tingle) needs to be inserted under the existing dpc. The tingle must be dressed over the top of the asphalt. The insertion needs to be carried out with extreme care. In most cases it will probably be impossible to insert the tingle without removing short lengths of the brick course above – no more than three bricks at a time. It may also be necessary to remove the asphalt in the chase for about 50 mm below it. The chase itself will probably need to be enlarged, taking care not to damage the dpc while so doing. After the tingle is inserted, the asphalt should be relaid into the chase, and after it has cooled, the tingle dressed over it. If, however, the upstand is a low parapet, it will be more economical to dismantle the whole of the parapet and rebuild it with a new dpc and apron flashing.

References for correct detailing
Duell and Lawson, *Damp-proof course detailing*, Fig. 22, emphasises the vulnerability of the dpc materials and therefore the need for special care.

Note: The details in the references listed elsewhere in this study all include the same principle, mainly of directing the water away from the interior of the building, but show the asphalt tuck-in and the dpc at different levels. They could usefully be consulted for guidance at least.

Manifestation 6
Internally: dampness at or near wall/ceiling junction.

Cause
Incorrect dpc/lead flashing relationship: either the dpc was stopped short of the lead or the lead was taken over (instead of under) the dpc, or the overlap was too short, allowing water to penetrate behind the asphalt, **14**.

Cross reference
Study 22: Rain penetration, Dpcs/parapets, 6.05, page 123.

Remedial work
1 *At best*: it may be possible to seal the gap or junction between dpc and lead with an appropriate sealant. This remedy will be short term (i.e. probable service life 5 to 10 years).
2 *At worst*: reform the overlap by taking out the lead flashing and inserting, with great care, a new one, so that the dpc overlaps it by at least 50 mm.
    (*a*) In a high parapet it may be possible to remove short lengths of the brick course above – no more than three bricks at a time – and probably the asphalt in the chase as well (the chase may have to be enlarged). After the lead is inserted, the asphalt should be relaid in the chase and after it has cooled, the lead dressed over it.
    (*b*) In a low parapet, it will be more economical to dismantle the parapet and rebuild with new dpc and lead flashing.
    (*c*) If appearance unimportant, capping the parapet and cladding the inside of the wall will be least disruptive and probably the most economical.

References for correct detailing
1 CP 144: Part 4: 1970, Fig 1, page 28 (see **6a**, page 122) shows the correct relationship and that the asphalt chase and the dpc/lead flashing should not be in the same brick course. If this is done, there is little chance of the asphalt being damaged, while the dpc/lead flashing can be built in as the work proceeds.
2 Duell and Lawson, *Damp-proof course detailing*, Fig. 22, emphasises the vulnerability of the dpc materials and therefore the need for special care.

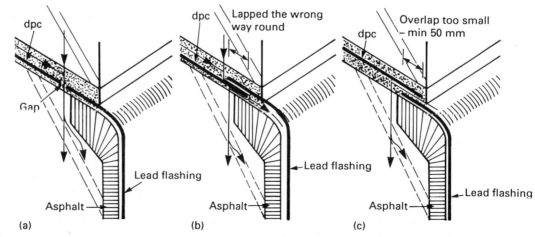

14 *Three conditions of the incorrect relationship between a dpc and lead flashing in the same joint as the top of an asphalt skirting*

Study 12
Pitched roofs

C/E 5

Manifestation 1
Dampness at ceiling level – not necessarily in any particular location but may be concentrated at the junction with the external wall – and/or timber decay of the trusses.

Condition
The schematic diagram **2**, shows the kind of constructional detail in which the above manifestation of failure may have arisen. It is a generalised detail, and the principle illustrated must be interpreted in relation to the specific construction found on site in any particular case.

Cause
1 Most commonly, condensation in the roof space of low-pitched roofs (i.e. roofs of about 15° pitch especially). Less commonly, driving rain or fine snow; see 4 in diagram **2**. The basic cause is due to the fact that the underlay is acting as a vapour control layer and is cold, so condensation forms underneath it (see 3 in diagram) and collects at the eaves or drops on to the ceiling.
Sufficiently high moisture content of air to allow condensation within the roof space is probably due to inadequate ventilation of the roof space at the eaves (see 5 in diagram) to remove the moisture penetrating into the roof space. This can occur either through the ceiling (see 1 in the diagram) which has inadequate vapour resistance (the latter is essential in low pitched roofs and desirable in high pitched roofs), or there is leakage of moisture laden air at gaps in the vapour check, around light drops and other projections through the ceiling and around trapdoors into the roof space (such leakage is probably the single most important source of moisture): and/or it can occur from the cavity (see 2 in the diagram) which should have been sealed at the top. The cavity can be an important source of moisture during the drying out of the building. In addition, insulated (and therefore reasonably warm) but not properly sealed tanks can be another source of moisture.
2 Timber decay, usually due to the dampness in the roof space and the fact that the timber was not preserved.

Cross references
Study 20: Condensation under heading 'Pitched roofs', 4.03, page 112.
Study 27: Timber decay, page 162.

Remedial work
1 *At best*, if there is no timber decay, and in this order:

   (a) Seal the top of the cavity.
   (b) Provide ventilation at the eaves by means of openings along two opposite sides equivalent to a continuous opening 10 mm wide for roofs above 15° pitch and 25 mm wide for roofs below 15° pitch.

60

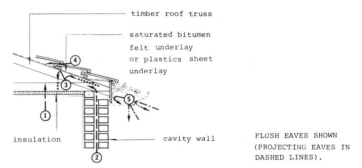

1 *Mould growth on underside of pitched roof, as a result of excessive moisture content in air and consequent condensation. (Photo BRE.)*

timber roof truss

saturated bitumen felt underlay or plastics sheet underlay

insulation   cavity wall

FLUSH EAVES SHOWN
(PROJECTING EAVES IN
DASHED LINES).

2 *Circled numerals indicate five common sources of high moisture content within roof space.*

Note: Low-pitched roofs require more ventilation because of the constriction at the eaves and because of the smaller volume of air in the roof space.
(c) Provide tightly sealed covers to water tanks.
(d) Where ventilation of the roof space and the room below do not prevent moisture from condensing in the roof space, a vapour control layer should be incorporated into the ceiling. Depending on the moisture conditions within the room below the ceiling, it may be sufficient to apply chlorinated rubber paint. Alternatively, PVC/foil-backed paper can be used. In all cases it is essential that all the gaps are properly filled with a sealant, particularly those around light fittings and other projections through the ceiling. In addition, trapdoors into the roof space should be 'weather stripped'.
2 *At worst*, if the timber is decayed, remove the decayed timber, treat it in close proximity with a preservative and renew the decayed portions with preserved timber.

References for correct detailing
1 BS 5250: 1989: *Code of Practice for Control of condensation in buildings*. A major revision of the 1975 Code. Overtakes in detail the references that follows and should therefore be used in preference to them. Clause 9.4.7, starting on page 19 deals with the many variants of pitched roofs:

9.4.7.1, page 19, Figures 9–11 gives guidance on the basic pitched roof form; 9.4.7.2, Figures 12–15, pages 20–21, inclined ceilings (i.e. no trussed rafters); 9.4.7.4, Figure 16, page 23, pitched roofs containing rooms with fully inclined ceilings; 9.4.7.5, Figure 17, pitched roofs containing rooms with partially inclined ceilings; and, 9.4.7.6, Figure 18, page 24, pitched roofs containing rooms with dormers.

2 Building Research Advisory Service Technical Information Leaflets 59, November 1977, *Condensation in domestic tiled pitched roofs – advice to householders,* puts in simple but practical terms the points made above about ventilating the roof space and reducing the sources of moisture in the roof space.

3 Handisyde, C. C., *Everyday details 22,* 'Pitched roofs: eaves', pages 107–109, discusses in simple text and clear illustrations the issues involved.

4 BRE DAS 1, *Slated or tiled, pitched roofs: ventilation to outside air* (Design). Principles and practice to provide adequate ventilation. Comprehensive coverage in BRE Digest 270.

5 BRE DAS 3, *Slated or tiled pitched roofs: restricting the entry of water vapour from the house.* (Design). Principles and practice. Comprehensive coverage in BRE Digest 270.

6 BRE DAS 4, *Pitched roofs: thermal insulation near the eaves* (Site). Methods to ensure adequate ventilation. Comprehensive coverage in BRE Digest 270.

7 BRE Digest 270, *Condensation in insulated domestic roofs.* More comprehensive and in some ways better than in BRE DASs above. Emphasises the importance of ventilation of the roof space and sealing around perforations through the ceiling leading into it rather than relying on a vapour control layer, the effectiveness of which is doubtful in any event. Has a useful checklist. BS5250: 1989 is more detailed.

Note: BRE has produced a short video, *Condensation in the home,* that covers the essential issues excellently (costs £20 from BRE Publications Sales Office).

8 BRE Report, *Thermal insulation:avoiding risks,* Pitched: Tile or slate, pages 6 and 7. Refers to BS5250 for size of ventilation gaps. Excellent drawings illustrating risky features including sealing of hatches, freezing of water in pipes and tanks and cold bridging. Text concise.

9 BRE DAS 16, *Walls and ceilings: remedying recurrent mould growth* (Design). Succinct on getting rid of mould – better covered in references 9 and 10 below.

10 BRAS, TIL 66, *Treatment of mould; advice to householders,* is good for its intended purpose. It is advisable to consult later references.

11 BRE Information Paper, IP 11/85, *Mould growth and its control,* has good basic coverage. It is advisable to consult later references.

12 BRE Digest 297, *Surface condensation and mould growth in traditionally-built dwellings.* Comprehensive coverage with selection of remedial measures and advice to occupiers. For the latter, *Condensation and mould growth in your home,* Domestic Energy Note 4 (DOE, HMSO, 1979), is 'lay' informative. (See note to BRE Digest 270 above.)

13 DOE, *Condensation in dwellings,* Part 2, Remedial measures (HMSO, 1971), is a little dated now but still has useful background.

14 BRE Digest 299, *Dry rot: its recognition and cure,* covers more than is normally required for this manifestation but could be useful in extreme cases of decay.

15 BRE Digest 345, Wet rots: recognition and control, complements BRE Digest 299: see above.

3 *Sagging of sarking at eaves with ponding. Water leaked at the unsealed laps. Photo also shows poorly executed cutting/sealing of a pipe projecting through a tiled roof.*

16 Richardson, Barry A., *Remedial treatments of buildings,* Chapter 2, Wood treatment, is comprehensive, including types of decay other than those encountered in this manifestation.

17 Desch, H. E., *Timber, its structure, properties and utilisation.* Chapter 15 for preservation of timber and Chapter 16 for the eradication of fungal and insect attack.

18 AJ Series, *Construction Risks and Remedies,* 'Condensation', 9 and 16.4.86, pages 49–58 and pages 69–81 has wide coverage that includes parts relevant to this manifestation.

19 AJ Series, *Construction Risks and Remedies,* 'Timber decay, 8.10.86 and 15.10.86. Comprehensive coverage on all aspects.

Manifestation 2
Dampness at ceiling level:
1 at or near the junction of the external wall
2 associated with perforations such as pipes or chimneys through the roof cladding.

Cause
Both are associated primarily with the sarking felt laid under the tiles or slate:

1 The sarking felt is not supported at the eaves to enable water to drain over the fascia into the gutter, so water ponds on the dip in the surface of the sarking behind the eaves fascia board **3**, and penetrates at unsealed laps and usually down the inner face of the inner leaf of the external cavity wall either directly or via the cavity, alternatively the sarking felt is not carried over bargeboards, so water can penetrate between the bargeboard and the wall.

2 The sarking felt is not properly fitted round the pipe, or chimney stack. An inadequate flashing on the external side of the cladding may make matters worse.

Cross reference
Study 22: Rain penetration, '7 Pitched roofs', page 124.

Remedial work
1 At the external wall:
   Eaves fascia: remove tiles or slates at eaves and at least one course up; redesign the eaves detail so that the sarking felt is properly supported by a tilting fillet or board with the top of the fascia board so related that

water can drain from the surface of the sarking into the gutter; lay narrow width of new sarking lapping *under* the existing by at least 150 mm; relay the tiles or slates. Bargeboard: remove sufficient tiles or slates to expose the sarking felt; extend the sarking felt with a new piece of felt lapped *under* the existing felt, supported over the cavity wall and carried over the top edge of the bargeboard; reinstate the tiles or slates.

2 At projections through the sarking: remove sufficient tiles or slates to expose the pipe (or chimney stack) and the sarking felt; patch the sarking felt with new felt sufficient in area to cover the existing felt for at least 300 mm around the pipe; the patch is to be bonded to the existing felt and is to have a cross cut in the centre for lowering over the pipe with the cuts projecting upwards; seal around the pipe; reinstate the tiles or slates including a pipe flashing, preferably of lead. (Around chimney stacks, the sarking felt should be sealed carried up the wall of the stack and sealed to it, ensuring that water from the back of the stack can drain down the sides of the stack.)

References for correct detailing
1 BRE DAS 9, *Pitched roofs: sarking felt underlay – drainage from the roof* (Design).
2 BRE DAS 10, *Pitched roofs: sarking felt underlay – watertightness* (Site).
3 BS 5534: Part 1: 1978, *Code of Practice for slating and tiling* (formerly CP 142: Part 2). Fig. 2 and clause 35.1, page 14, for eaves support of sarking felt at eaves. The code is silent on what should be done to the sarking felt around projections through the roof; clause 44, page 16, deals with flashings only.

Manifestation 3
Dampness at ceiling level or leaks from ceiling.

Condition
The schematic section and detail in diagram **4** shows the kind of constructional detail for plastic-coated profiled metal (usually steel) roof cladding in which the above manifestation of failure may have arisen. It is a generalised section and detail, and the principle illustrated must be interpreted in relation to the specific construction found on site in any particular case.

Cause
Usually interrelated or interdependent, and therefore one or more of:
1 Pitch too low for exposure to wind and/or driving rain. Normal structural deflections may make matters worse.
   (a) end lap too small for exposure to driving rain;
   (b) not fully sealed (**5**);
   (c) fasteners incorrectly located (**6**) (affects degree to which sealant is compressed), or screwed in skew or overtightened with or without enlargement of hole (distorts compression washer and reduces its effectiveness to exclude rainwater or may affect degree to which sealant is compressed or may crush compressible insulation below).
3 Side lap:
   (a) faces in the direction of the prevailing rain;
   (b) not fully sealed (**7**);
   (c) side lap stitching omitted, too widely spaced or incorrectly fixed.

4 Fillers omitted or wrongly installed.
5 Rooflights:
   (a) mismatch of profiles;
   (b) End laps between sheets or between sheets and metal not properly fixed or sealed (**8**);
   (c) Distortion of sheets due to mishandling and/or thermal/ultraviolet effects of the sun;
   (d) Seal of 'double-glazed' units poorly made or broken.
6 Gutter not fully supported and/or downpipes are blocked.
7 Interstitial condensation mainly because vapour control layer is ineffective and/or because of external ventilation air between the top of the insulation and the hollow of the profile.
8 Thermal pumping at end and/or side laps (usually at unsealed or poorly sealed laps, but the action may even break good seals) by which external air is drawn into or driven out of the cavities above the insulation. Moist air entrapped within the cavities can subsequently condense – a newly recognised process.

Cross references
1 Study 20: Condensation, under heading 'Pitched roofs', 4.03, page 112
2 Study 22: Rain penetration, under heading '7 Pitched roofs', page 124

Remedial work
Note: Unlike the other manifestations in this and other studies, it is not possible to give definitive guidance on the nature and scope of remedial work that may be necessary. The scope of work depends on the nature and extent of the defects that caused the failure. If it is necessary to remove fasteners in order to remedy a defect in the fasteners or in the laps, for example, it is usually not possible to refix the fasteners properly unless larger-diameter fasteners are used. (The fasteners are usually self-tapping, and these cannot be 'rescrewed' successfully.) For much the same reason, replacing sheets where the defects have been shown to occur is also seldom entirely successful. In most cases it is usually necessary to replace the sheeting to the whole of the roof, often using different design principles or details.

1 Where the defects are shown to be local and not widely distributed, these may be remedied as appropriate.
2 It is more likely to be the case that the whole of the existing roofing will need to be stripped and relaid with new roofing, incorporating different design principles or details. Where there are rooflights, these are preferably not reinstated or reinstated at or near the ridge. Rooflights are better avoided.
3 An alternative to stripping the existing roofing is (subject to the structure being strong enough) to convert the roof to a warm deck design by using the existing profiled cladding as a roof deck and superimposing a vapour control layer (to help with wind uplift if not required to deal with water vapour) bonded to the existing cladding, board insulation (of sufficient thickness to 'counteract' the value of any existing insulation) bonded to the vapour control layer and then laying built-up felt, preferably high-performance felt.

References for correct detailing
Note: There is as yet no authoritative guidance by way of BS codes of practice. BS 5427 and CP 143 (in many parts) listed below do not cover recent developments. BS 5427: 1976 does however contain the essential principles and

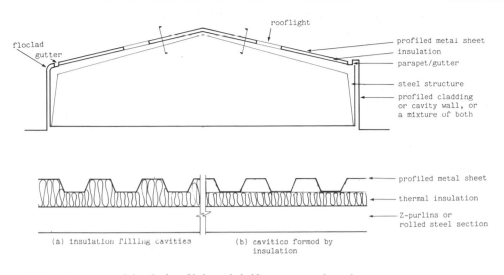

4 *Schematic section and detail of profiled metal cladding as commonly used.*

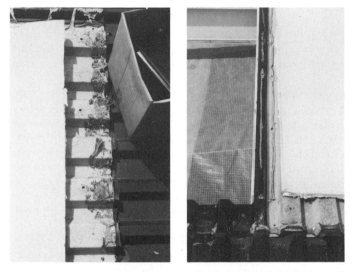

5 *underside of end lap of profiled steel sheet – plenty of sealant but little contact with sheet below.*
6 *Four fasteners per sheet width recommended by maker; three fixed.*

8 *Ill-fitting profiles between rooflight sheets (centre) and rooflights and metal – top and bottom laps.*

7 *Side lap not fully sealed. End lap – lower part of photograph the same.*

other useful design data. The other references are selective and give reasonably sound guidance, bearing in mind that thinking on the problem is subject to change as the number of cases of failures investigated increases. References to manufacturer's technical literature has been avoided. When such literature is consulted, it should be borne in mind that some recommendations, in particular on shallow pitches, are not necessarily based on research or experience.

1 BS 5427: 1976, *Code of practice for performance and loading criteria for profiled sheting in building.*
2 CP 143: *Code of practice for sheet roof and wall coverings.* Parts 1: 1958 and 10: 1973 cover corrugated aluminium and steel respectively.

3 National Federation of Roofing Contractors, *Profiled sheet metal roofing and cladding, a guide to good practice* provides basic guidance but does not address problems such as thermal pumping.

4 Peter Falconer, *Metal industrial roofs, moisture problems, AJ,* 7.05.86, pages 53–55. Helpful analysis of the main problems and optimism of manufacturers. Discusses mechanics of secondary interstitial condensation (i.e. condensation in the cavities above the insulation from external ventilation air).

5 *AJ* Series, *Element Design Guide: Roofs,* '4 Profiled sheet roofs' by Peter Falconer. Profiled metal roofs dealt with comprehensively – more so than earlier article (2 above) – and is packed with details.

6 *AJ* Series, *Construction Risk and Remedies,* 'Condensation', Part 1 (9.04.86), section 5.2 (page 58) and Part 2 (16.04.86), section 4.3 (page 16) and 'Avoiding sheet roof problems' (page 79), covers some of the important issues, and is good for background reading.

7 PSA, 'Lap joints in low pitch profiled roofs', Building File, 17, April 1987, page 15, discusses and examines briefly the performance of various design of lap joints.

8 BRE DAS 59, *Felted cold deck roofs: remedying condensation by converting to warm deck* (Design), is included here for principles to be applied if existing roof is to be converted to a warm deck design by using the existing profiled metal as a deck.

Manifestation 4
Cracking of asbestos cement or fibre cement corrugated sheeting.

Cause
1 Hole in sheet for fastener too small and/or fastener overtightened resulting in the sheets being restrained, usually from moisture movements and then cracking either along the crown having the fasteners or along a crown at mid-sheet width, the latter consequent on the sheet bowing.

2 One-sided carbonation of externally painted sheets, the carbonation causing shrinkage on the internal face of the sheet. Restraint from movements as in 1 above may help to make the cracking worse.

Cross reference
Study 24: Movement, under '3 Types of movement', page 134 and '4 Causes of failure', page 136.

Remedial work
1 For fastener defects: replace affected sheets taking care if self-tapping fasteners were used that the new fasteners are larger than those being replaced.

2 For carbonation defects: replace all sheets. If replacement sheets are to be painted externally, ensure that there is also a balancing coat of paint on the inside face.

References for correct detailing
1 BS 5247: Part 14: 1975: *Code of Practice for sheet roof and wall coverings, corrugated asbestos-cement,* Clause 28, page 10 for fixing.

2 BRS Digest 130 (First series), *Asbestos cement sheets cracking,* for need to paint both sides. Also in *Principles of modern building,* Vol. 2, page 166.

Manifestation 5
Loosening of fasteners in profiled sheet roofs of asbestos cement, cement fibre or metal, **10**. Enlarging of hole for

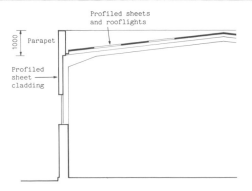

9 *Parapet conditions that can give rise to unusual wind effects (e.g. vortices).* **a** *Diagrammatic section.* **b** *Photographic example of the condition shown in* **a**. *Note bending of vent and its stays; in background, sheets are loose. All the effects of unusual wind effects, vortices in particular. Building sited in a windy part of the country.*

10 *Enlargement of fixing holes in asbestos cement sheeting due to fretting of the asbestos cement as a result of 'chattering' by the wind due to high parapet. From example in* **13b**.

fastener in asbestos cement sheets (could also occur in cement fibre sheets). Also loosening of parapet capping and sheets behind the parapet, if the parapet is a continuation of profiled wall cladding, particularly if of asbestos cement.

Condition
The schematic section in **9** shows the kind of parapet/roof relationship in which the above manifestation of failure may have arisen. It is a generalised section, and the principle illustrated and discussed below must be interpreted in relation to the specific construction found on site in any particular case.

Cause
Basically, vortex or vortex-type action of the wind because of high parapets (one metre or higher) around a roof with low pitches and notably in locations severely exposed to high winds:

1 On the roof side at or near the parapet: the speed of the wind is increased and is accompanied by vibrations that tend to cause asbestos sheets to 'chatter'. Continual exposure to this action causes the asbestos around the fastener hole to 'fret', thereby loosening the fastener.

2 Elsewhere on the roof: the vortex action, and the vibrations with it, tend to 'bend' the sheets between fasteners. In time the fasteners unscrew. (The vortex action is increased if the roof behind the parapet has a sawtooth profile.)

Cross reference
Study 24: Movements, page 134.

Remedial work
It may be possible to replace the fixings with alternatives that are better able to withstand the wind effects, but the long-term solution lies in reducing the height of the parapet and, if the condition of the sheeting is generally sound, to replace the affected fasteners with new ones of a wider diameter. Damaged sheets should be replaced as required.

References for correct detailing
None. Consult wind specialists if high parapets are essential.

Study 13
Cavity walls

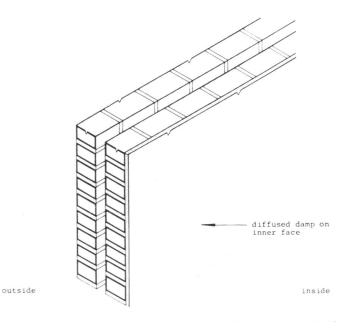

Manifestation 1: Dampness – condensation

Dampness on the inside surface of the wall, usually with mould growth. The dampness is *not* associated with rain (see diagnostic check list in Studies 20 and 22. See also Study 3, 3.02 for clues).

Cause
Surface condensation due to:

1 Inadequate rate and inappropriate mode of ventilation, in relation to the amount of moisture generated inside the building.
2 Inadequate thermal insulation of the wall (localised areas of dampness could be due to cold bridges: see Study 16: window openings).

3 Inappropriate thermal capacity of the wall.
4 The thermal capacity of the heating system and/or its use being too intermittent, thus allowing the walls to become excessively cold between heating periods.

Cross references
Study 20: Condensation under heading 'Surface condensation', 4.02, page 108.
Study 22: Rain penetration under heading 'Cavity walls', 8, page 126.

Remedial work
In principle:

1 Improve the ventilation; and/or
2 Improve the thermal insulation of the wall with additional insulation applied externally, internally or in the cavity; and/or
3 Change the running pattern of the heating system; and/or
4 Provide additional heating.

Adequate ventilation is required to prevent an excessive build-up of moisture in the internal air, particularly in rooms where there is a high production of moisture such as in kitchens and bathrooms; swimming pools or indoor sports halls; or, from industrial processes. When heating is intermittent, walls of high thermal capacity (brick or concrete) are slow to warm up and may be cold enough when the heating recommences to allow surface condensation. Low levels of heating for longer periods will help to keep the walls at a more constant temperature.

1 *Manifestation of damp, of kind which is probably not associated with rain penetration. Diffuseness is one clue (compare with well-defined patches in* **3a, b**); *presence of mould (which requires relative humidity of air to be sustained above 70 per cent to allow growth) is another. But hypothesis needs to be tested before conclusion is drawn — see checklists in Study 20, pages 108ff.*

diffused damp on inner face

outside   inside

2 *Surface condensation can occur with many basic wall types, particularly if thermal insulation value is inadequate.*

References for correct detailing

1 General:

   (a) DOE/HMSO: *Condensation in dwellings Parts 1 and 2.* In Part 2, pages 25–28 give general principles for estimation of condensation in particular rooms. Pages 22–29 of the same part describe in detail remedial measures.

3 *Well-defined patches of damp probably due to rain penetrating the wall (as opposed to diffuse damp shown in* **1,** *which is probably due to condensation). The specific underlying cause is most commonly one of those illustrated in conditions* **1, 2** *or* **3** *on this page and pages 68, 70. (***3a** *by Building Research Establishment, Crown Copyright.)*

   (b) BS 5250: 1989 *British Standard Code of Practice for Control of condensation in buildings,* 4, 5, 6, 7 and 8, pages 6–11.

   (c) BRE Report *Thermal insulation: avoiding risks,* 2 Walls, pages 12–19, covers concisely yet comprehensively all the key issues not only related to condensation risk but other aspects.

   (d) BRE Report, *Rain penetration through masonry walls: diagnosis and remedial measures,* 1988. Excellent coverage. Most useful.

   (e) BRE IP 12/88, *Summer condensation on vapour checks: tests with battened, internally insulated, solid walls,* September 1988. Timely caution when insulating internally see BRE DAS 133 below for application.

   (f) BRE DAS 133, *Solid external walls: internal dry-lining – preventing summer condensation,* June 1989, applies the nuts and bolts of BRE IP 12/88 above.

2 For cavity insulation:

   (a) BRE Digest 236: *Cavity insulation,* discusses the various forms of insulant available and recommends using blown-in rock fibre or polystyrene beads for least risk of rain penetration.

Note: The blown-in insulants have the added advantage that they can at least be taken out of the cavity, albeit mostly with some difficulty, if remedial work has to be carried out within the cavity to overcome problems of rain penetration (13).

   (b) BS 5618: 1985: *Code of Practice for the thermal insulation of cavity walls (with masonry or concrete inner and outer leaves) by filling with urea–formaldehyde foam systems* is a new edition of the code that updates suitable properties and tightens the installation procedure and supersedes the 1978 edition. It is likely to be overtaken if and when DD 93 becomes a code of practice – see (c) below.

   (c) DD 93: 1984: *Draft for development, Methods of assessing exposure to wind-driven rain,* adopts completely different and more 'accurate' methods – see Study 22, Rain penetration, page 120. It is likely to be overtaken if and when DD 93 becomes a code of practice and is already referred to in BS 5628: Part 3: 1985 – see (g) below.

   (d) Agrément Certificates for individual products (at least 75 certificates by mid 1989).

Note: Most, if not all, Agrément Certificates give the limitations of the product or the circumstances of its use. It is important that these are read and interpreted for particular cases. (See also Information sheets below.)

   (e) BBA, Information sheet 10, *Methods of assessing the exposure of buildings for cavity fill insulation* (revised edition, January 1983) relates to and formed the basis of BS 5618. (See reference (c) above.)

   (f) BBA, Information sheet 16, *Cavity wall insulation,* has important qualifications as to the suitability of certain walls for cavity fill, including the interpretation of local practice, and for severe exposure conditions the increased risk of water penetration when low-porosity bricks and/or recessed joints are used. These qualifications were seldom included in certificates for particular products.

   (g) BS 5628: Part 3: 1985: *British Standard Code of Practice for use of masonry,* 'Materials and components, design and workmanship' (formerly CP 121: Part 1: 1973). Table 11, pages 38 and 39, includes cavity fill as one of the factors affecting the rain penetration of cavity walls (clause 21.3.2.8, page 39, relates).

Note: For classification of exposure to wind-driven rain, the code uses a 'local spell index' as described in DD 93, which is compared with exposure categories in CP 121: Part 1: 1973 (Table 10, page 36).

(h) *AJ* Series, *Construction Risks and Remedies*, 'Condensation', 9.04.86, pages 49–58, and 16.04.86, pages 69–81, is useful for background and has a comprehensive list of references.

(i) *AJ* Series, *Construction Risks and Remedies*, 'Thermal insulation', 11.06.86, 3 Cavity walls, pages 58 and 60, 18.06.86, 5 Cavity wall remedies pages 71 and 73, is useful as background and has a comprehensive list of references.

3 For externally applied insulation:

(a) GLC, *Development and Materials Bulletin*, **129**, October 1980, item 5 'External insulation systems', describes three basic systems:

- Rendering using expanded polystyrene beads in a cementitious matrix;
- Expanded polystyrene boards fixed to the outside of the building and protected by a decorative finish;
- Composite boards made of glass fibre or mineral wool batts with interwoven steel mesh and breatherpaper fixed to them. A number of proprietary systems are described in more detail. Followed up in 131, item 5 – see below.

(b) GLC, *Development and Materials Bulletin*, **131**, February 1980, item 5 'External insulation systems 2', is a follow-up of **129**, item 5 (above). Additional systems examined.

(c) GLC, *Development and Materials Bulletin*, **139**, item 3 (Approved: 6/83), *External insulation systems for walls – compliance with GLC requirements*, Building Technical File, 3, October 1983, page 13. Read in conjunction with GLC *Bulletins* **129** and **131**–(a) and (b) above.

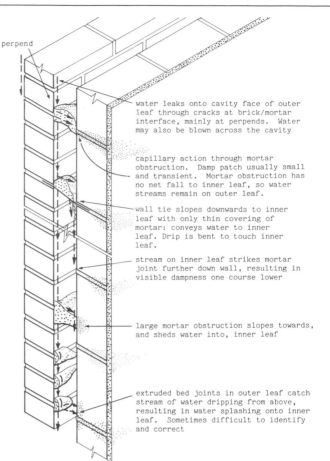

perpend

water leaks onto cavity face of outer leaf through cracks at brick/mortar interface, mainly at perpends. Water may also be blown across the cavity

capillary action through mortar obstruction. Damp patch usually small and transient. Mortar obstruction has no net fall to inner leaf, so water streams remain on outer leaf.

wall tie slopes downwards to inner leaf with only thin covering of mortar: conveys water to inner leaf. Drip is bent to touch inner leaf.

stream on inner leaf strikes mortar joint further down wall, resulting in visible dampness one course lower

large mortar obstruction slopes towards, and sheds water into, inner leaf

extruded bed joints in outer leaf catch stream of water dripping from above, resulting in water splashing onto inner leaf. Sometimes difficult to identify and correct

4 *Forms of rain penetration (see Manifestation 2) due to inadequacy of walling material, and/or faulty cavity ties. Based on 'Cavity insulation of masonry walls – dampness risks and how to avoid them'*

Manifestation 2: Dampness – rainwater

Dampness on the insides of the wall; the dampness is associated with rain (see diagnostic checklists in Studies 20 and 22. See also Study 3, 3.02 for clues).

Notes:

1 This manifestation includes cavities that are unfilled, filled or partially filled. There are five different conditions. Each is self-contained except for the 'References for correct detailing'. These have been gathered together and are given after Condition 5. For convenience, they are grouped.

2 When considering 'Causes' and 'Remedial work' under each Condition, note that filling the cavity has the effect of highlighting or revealing faults that in an unfilled cavity may not cause problems.

Condition 1: Unfilled cavity – faults, **4**

Cause

1 Outer leaf:

(a) The bricks or blocks are too permeable (notably some types of lightweight concrete blocks) for the degree of exposure to driving rain; or

(b) The bricks are too dense and/or too impermeable or absorb too little water for the degree of exposure to driving rain. The problems can be and usually are aggravated if jointing is recessed, **5**.

5 *Dense (i.e. low-absorbency) bricks with recessed joints.*

8 *Mortar extrusions on the inner face of the outer leaf help to direct water across the cavity.*

6 *Inadequate jointing:* **a** *cracks/throughway in jointing (perpends);* **b** *faces of headers showing non-contact of mortar with the brick surface although the perpends appeared to be filled with mortar.*

9 *Dirty cavity tie – directs water to inner leaf.*

7 *The (high) perpends of soldier courses are difficult to fill properly.*

(c) In both (a) and (b) the bed joints and/or perpends are not properly filled, **6**, and/or there are small cracks between the mortar and the masonry unit.

Note: In soldier course brickwork it is not uncommon to find that the perpends are not properly filled. Filling such joints needs care and skill in laying the bricks, **7**.

(d) There are mortar extrusions (snotting) on the inside face of the leaf, **8**.

2 Water may be blown across the cavity inner leaf because of causes in 1 above but, in particular, (c) and (d).

3 Water may bridge across the cavity because of dirty or sloping or badly located wall ties, **9**.

4 If the damp occurs near or associated with rc beams or nibs, see Study 14: Manifestations 4 and 5, page 81, for bridging at faulty cavity trays or dpcs.

5 If the damp occurs near openings, see Study 16: Window openings, page 85, for bridging at faulty cavity trays or dpcs.

Cross reference
Study 22: Rain penetration under heading '8 Cavity walls', page 126.

Remedial work

At best:

1 Treat the wall with a colourless water-repellent externally (with caution and demanding impeccable application/
supervision).

   Note: There is a body of opinion, based on laboratory
   experiments and being confirmed in practice, that
   considers that the jointing in the outer leaf takes time to
   mature and for small cracks and fissures to heal, the
   process described as 'self-healing'. Consequently the
   leakiness of newly built leaves may be temporary.

2 Rake out affected joints to a depth of about 20 mm and
repoint. Ensure that small areas only are remedied at a
time, more so if the joints are to be raked out to a depth of
greater than 20 mm.

At worst:

3 For correcting localised badly unfilled joints or to clear
the cavity of blockages (e.g. dirty ties or bad snots):
remove the related bricks but in small areas only and
then reinstate (after clearing blockages).

   Note: It is rarely possible to ensure that the jointing of the
   last bricks reinstated will be filled perfectly. Such
   imperfection may be important, more so if the cavity is to
   be filled in the future.

4 For correcting widespread defects of all kinds: render or
clad the wall externally.

References for correct detailing
At the end of Condition 5 – see Manifestation 2, Note 2.

Condition 2: Unfilled cavity – failure

Cause

1 Loss of adhesion; or cracking of mortar joints (mainly
horizontal); or cracking of rendering due to movements, allowing penetration of water through the outer
leaf. This water may then be blown across the cavity, or
bridge across dirty or sloping ties, to reach the inner leaf.
Loss of adhesion or cracking, as described above, will
allow rain penetration if the former is excessive for the
degree of exposure to driving rain. The latter therefore,
should be taken into account in diagnosis/remedial work.

2 If the damp occurs near or is associated with rc beams or
nibs, see Study 14: Manifestations 4 and 5, page 81, for
bridging at faulty cavity trays or dpcs.

3 If the damp occurs near openings, see Study 16: Window
openings for briding at faulty cavity trays and dpcs.

Cross references
Study 25: Loss of adhesion, under headings '4 Mortars',
page 147, and 'External render', 7, page 151.
Study 22: Rain penetration, under heading '8 Cavity walls',
7, page 126.

References for correct detailing
At the end of Condition 5 – see Note 2 under Manifestation
2.

Condition 3: Filled cavity – foam

Cause

1 The use of (foam) cavity filling in conditions of
excessively severe degree of exposure to driving rain,
resulting in bridging across the filling – at breaks in the
filling or around dirty or sloping ties.

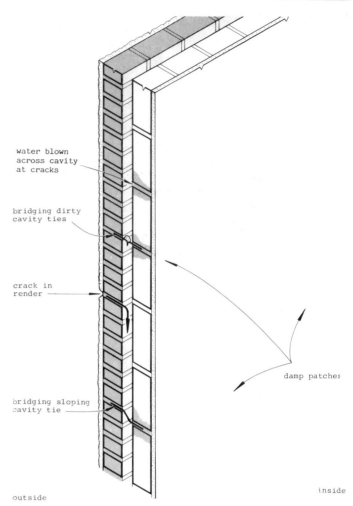

water blown
across cavity
at cracks

bridging dirty
cavity ties

crack in
render

bridging sloping
cavity tie

damp patches

outside   inside

10 *Forms of rain penetration (see Manifestation 2) due to cracks in wall or
finish, and faulty cavity ties. (Routes similar to **4**.)*

2 If the damp occurs near or is associated with rc beams or
nibs, see Study 14: Manifestations 4 and 5, page 80, for
bridging at faulty cavity trays or dpcs.

3 If the damp occurs near openings, see Study 16: Window
openings, page 85, for bridging at faulty cavity trays and
dpcs.

Cross reference
Study 22: Rain penetration under heading 'mortar mix,
render mix and cavity fill', 8.03 page 127.

Remedial work

1 Treat the wall with a colourless water-repellent externally (with caution and demanding impeccable application/
supervision).

   Note: There is a body of opinion, based on laboratory
   experiments and being confirmed in practice, that
   considers that the jointing in the outer leaf takes time to
   mature and for small cracks and fissures to heal, the
   process described as 'self-healing'. Consequently the
   leakiness of newly built leaves may be temporary.

2 Render or clad the wall externally.

References for correct detailing
At the end of Condition 5 – see Note 2 under Manifestation
2.

Condition 4: Filled cavity – not foam, 12
Note: When considering 'Causes' and 'Remedial work' under this condition, note that filling the cavity has the effect of highlighting or revealing faults that in an unfilled cavity may not cause problems.

Cause
1 Outer leaf:
  (a) The bricks or blocks are too permeable (notably some types of lightweight concrete blocks) for the degree of exposure to driving rain; or
  (b) The bricks are too dense and/or too impermeable or absorb too little water for the degree of exposure to driving rain. The problems can be (and usually are) aggravated if the jointing is recessed, **5**.
  (c) In both (a) and (b) the bed joints and/or perpends are not properly filled, **6**, and/or there are small cracks between the mortar and the masonry units.

Note: In soldier course brickwork it is not uncommon to find that the perpends are not properly filled. Filling such joints needs care and skill in laying the bricks, **7**.

  (d) There are mortar extrusions (snotting) on the inside face of the leaf, **8**.
2 Water finds its way across the cavity to the inner leaf because of causes in 1 above but in particular (c) and (d).
3 Water may bridge across the cavity because of dirty or sloping or badly located wall ties, **9**.

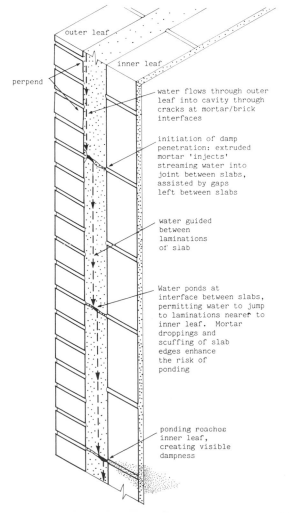

11 *Rain penetration (see Manifestation 2) due to faulty or inappropriate inside cavity fill. Based on 'Cavity insulation of masonry walls – dampness risks and how to avoid them'*

4 The fill – loose or loose bonded:
  (a) There are spalls of brick or concrete (surface of either facing the cavity 'chipped' off when drilling to inject insulation material) bridging the cavity.
  (b) For fills with a bonding agent – the bonding agent has not cured fully. (Most of the bonding agents are water based. Until the water has dried out, the wet material allows water to be transferred across it with comparative ease.)
5 The fill – slabs or batts:
  (a) The laminations are wrongly orientated.
  (b) There are gaps at the joints between the slabs or batts. (These may be filled with mortar extrusions or dirty wall ties. Either aggravates the problem.)
  (c) Compression of insulating material by mortar extrusions.
6 If the damp occurs near or is associated with rc beams or nibs, see Study 14: Manifestations 4 and 5, pages 81, for bridging at faulty cavity trays or dpcs.
7 If the damp occurs near openings, see Study 16: Window openings, pages 85–91 for bridging at faulty cavity trays or dpcs.

Cross reference
Study 22: Rain penetration under heading '8 Cavity walls', page 126.

Remedial work
At best:
1 Treat the wall with a colourless water-repellent externally (with caution and demanding impeccable application/supervision).

Note: There is a body of opinion, based on laboratory experiments and being confirmed in practice, that considers that the jointing in the outer leaf takes time to mature and for small cracks and fissures to heal, the process described as 'self-healing'. Consequently the leakiness of newly built leaves may be temporary.

2 Rake out affected joints to a depth of about 20 mm and repoint. Ensure that small areas only are remedied at a time, more so if the joints are to be raked out to a depth of greater than 20 mm.

At worst:
3 For correcting localised badly unfilled joints or to clear the cavity of blockages (e.g. dirty ties or bad snots): remove the related bricks but in small areas only and then reinstate (after clearing blockages).

Note: It is rarely possible to ensure that the jointing of the last bricks reinstated will be filled perfectly. Such imperfection may be important, more so if the cavity is to be filled in the future.

4 For correcting widespread defects of all kinds: render or clad the wall externally.

Note: Overcladding of the wall should provide a satisfactory solution even in the worst cases. Nevertheless, there have been instances where demolition of the outer leaf and rebuilding it, sometimes with a wider cavity so that insulation in a partially filled cavity may be incorporated, has been carried out. As draconian a measure as this is often proposed in litigation. I would find such a measure hard to justify technically or economically.

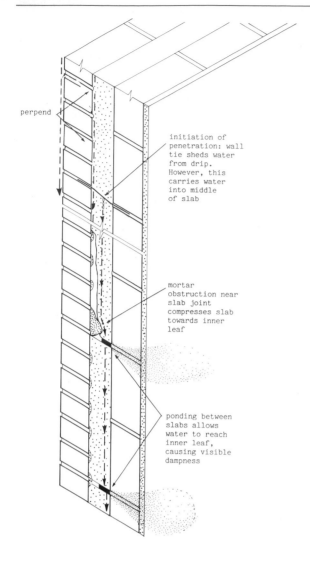

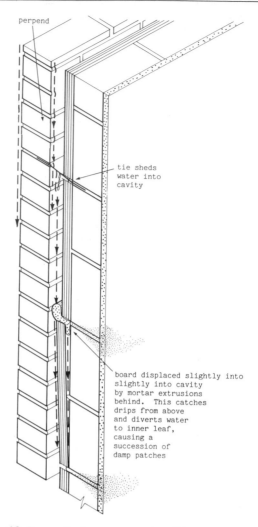

12 *Forms of rainwater penetration (Manifestation 2, Condition 5) due to inadequacy of walling material and/or faulty cavity ties and/or incorrectly installed board cavity fill. Based on 'Cavity insulation of masonry walls – dampness risks and how to avoid them'*

References for correct detailing
At the end of Condition 5– see Note 2 under Manifestation 2.

Condition 5: Partially filled cavity, 13

Cause
1 Outer leaf:
   (a) The bricks or blocks are too permeable (notably some types of lightweight concrete blocks) for the degree of exposure to driving rain; or
   (b) The bricks are too dense and/or too impermeable or absorb too little water for the degree of exposure to driving rain. The problems can be (and usually are) aggravated if the jointing is recessed, **5**.
   (c) In both (a) and (b) the bed joints and/or perpends are not properly filled, **6**, and/or there are small cracks between the mortar and the masonry units.

Note: In soldier course brickwork it is not uncommon to find that the perpends are not properly filled. Filling such joints needs care and skill in laying the bricks, **7**.

   (d) There are mortar extrusions (snotting) on the inside face of the outer or inner leaf, **8**.
2 Water finds its way across the cavity to the inner leaf because of causes in 1 above but in particular (c) and (d).

3 Water may bridge across the cavity because of dirty or sloping or badly located wall ties, **9**.
4 Board, slab or batt:
   (a) Displacement by mortar so that board projects into the cavity.
   (b) Gap at joints between board filled with mortar.
   (c) Laminated board wrongly orientated.
5 If the damp occurs near or is associated with rc beams or nibs, see Study 14: Manifestations 4 and 5, page 81, for bridging at faulty cavity trays or dpcs.
6 If the damp occurs near openings, see Study 16: Window openings, page 85, for bridging at faulty cavity trays or dpcs.

Cross reference
Study 22: Rain penetration under heading '8 Cavity walls', page 126.

Remedial work
At best:
1 Treat the wall with a colourless water-repellent externally (with caution and demanding impeccable application/supervision).

Note: There is a body of opinion, based on laboratory experiments and being confirmed in practice, that

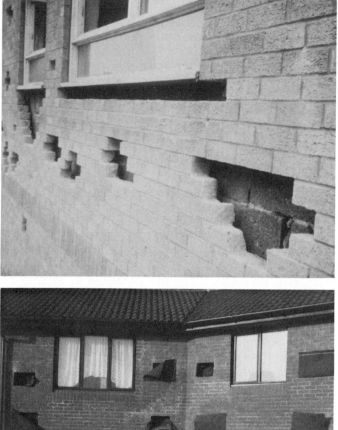

13 *Two examples of what might be involved in removing blown-in cavity fill:* **a** *fibreglass and* **b** *bonded expanded polystyrene beads.*

considers that the jointing in the outer leaf takes time to mature and for small cracks and fissures to heal, the process described as 'self-healing'. Consequently the leakiness of newly built leaves may be temporary.

2 Rake out affected joints to a depth of about 20 mm and repoint. Ensure that small areas only are remedied at a time, more so if the joints are to be raked out to a depth of greater than 20 mm.

At worst:

3 For correcting localised badly unfilled joints or to clear the cavity of blockages (e.g. dirty ties or bad snots): remove the related bricks but in small areas only and then reinstate (after clearing blockages).

Note: It is rarely possible to ensure that the jointing of the last bricks reinstated will be filled perfectly. Such imperfection may be important in some cases.

4 For correcting widespread defects of all kinds: render or clad the wall externally.

References for correct detailing (Conditions 1–5)
1 General:
(a) *Water off a duck's back (or how not to sink like a brick)*, by Donald Foster, Building Design Brick Supplement, 1983. Do not be put off by the title. It addresses rain resistance of cavity walls and is one of the most lucid explanations of the factors involved, including the effects of filling the cavity.
(b) BRE Report *Thermal insulation: avoiding risks*, 2 Walls, pages 12–19, covers concisely yet comprehensive all thekey issues not only related to condensation risk but other aspects such as resistance to rain penetration. Sensible use of tables and well illustrated.
(c) BS 3921: 1985: *Specification for clay bricks* is a revised standard that has a new classification for clay bricks, with particular relevance to the specification of durable bricks.
(d) BS 5250: 1989: *Code of Practice for control of condensation in buildings.*
(e) BS 5628: Part 3: 1985: *British Standard Code of Practice for use of masonry*, 'Materials and components, design and workmanship' (formerly CP 121: Part 1: 1973).

Note: For classification of exposure to wind-driven rain, the code uses a 'local spell index' as described in DD 93, which is compared with exposure categories in CP 121: Part 1: 1973 (Table 10, page 36). It covers a wide range of aspects, including brief reference to cavity fill.

(f) DD 93: 1984: *Draft for development, Methods of assessing exposure to wind-driven rain*, adopts completely different and more 'accurate' methods – see Study 22, Rain penetration, page 118. It is likely to be overtaken if and when DD 93 becomes a code of practice and is already referred to in BS 5628: Part 3: 1985.
(g) *AJ* Series, *Element Design Guide*, 'External walls 1 Masonry: Part 2', by John Duell, 9.07.86, 'Thermal and moisture', pages 45–47. Basic coverage with emphasis on the requirements of the Building Regulations and a good list of references.
(h) *AJ* Series, *Construction Risks and Remedies*, 'Condensation', 9.04.86, pages 49–58 and 10.04.86, pages 69–81, is useful for background and has a comprehensive list of references.
(i) *AJ* Series, *Construction Risks and Remedies*, 'Thermal insulation', 11.06.86, 3 Cavity walls, pages 58 and 60, 18.06.86, 5 Cavity wall remedies, pages 71 and 73, is useful as background and has a comprehensive list of references.
(j) BRE Report *Rain penetration through masonry walls: diagnosis and remedial measures*, 1988.
2 For liquid treatments:
(a) BS 3826: 1969: *Silicone-based water repellents for masonry* gives the requirements for the application and performance of the water repellants. Appendix H gives notes on their use.
(b) BS 6477: 1984: *Specification for water repellents for masonry surfaces* gives performance requirements and tests. Appendix gives recommendations for application.
(c) GLC, *Development and Materials Bulletin*, 129, October 1980, item 4 'Brickwork waterproofing solutions' describes tests carried out on 45 products for their resistance to water penetration, resistance to efflorescence and ability to 'breathe', i.e. to alllow the transfer of water vapour. Sixteen products were recommended. The report warns that if the internal dampness is due to condensation, faulty dpcs or

defective mortar joints, then the application of a silicone solution may either give no benefit or aggravate the conditions. Recoating may be necessary at frequent intervals.

(d) Agrément Certificates for water-repellant coatings.

3 For repointing/repairing brickwork:

(a) BRE Digest 160, *Mortars for bricklaying*, relates to CP 121: Part 1: 1973, now superseded by BS 5628: Part 2: 1985. The principles remain unaltered by new code and present mortar mixes differently and with fewer categories.

(b) BRE Digest 200, *Repairing brickwork*, has wide-ranging coverage.

(c) BRE DAS 70, *External masonry walls: eroding mortar – repoint or rebuild?*, for diagnosis and remedial.

(d) BRE DAS 70, *External masonry walls: repointing – specification* (Design).

(e) BRE DAS 70, *External masonry walls: repointing* (Site).

(f) Nash, W. G., *Brickwork repair and restoration*, addresses all the important issues with plenty of illustrations of terms, constructions and details. It is good practical stuff with no references, and is probably more useful for repairing or remedying pre-war buildings.

(g) DOE, *Construction* **21**, March 1977, page 26, 'Rain penetration of brickwork'. Simulated rain penetration tests showed that mortars containing lime (1:1:6 cement:lime:sand) have the greatest resistance to absorption.

(h) BRAS, *Warehouse cavity walls: defects causing rain penetration*, Building Technical File, 15, October 1986, pages 20–23, is useful for repointing.

4 For rendering:

(a) BS 5262: 1976: *External rendered finishes* (formerly CP 221).

(b) C&CA Series, *Appearance Matters* 2, 'External rendering', relates to and discusses in simpler terms BS 5262: 1976 Code of practice for external rendered finishes.

(c) BRE Digest 196, *External rendered finishes*, relates to and explains BS 5262: 1976.

(d) BRE DAS 37, *External walls: rendering – resisting rain penetration* (Design), specifies mixes and advises on precautions.

(e) BRE DAS 38, *External walls: rendering – resisting rain penetration* (Site), guidance on practical measures.

5 For cladding:

Note: Cladding may consist of 'traditional' tile or slate hanging or weatherboarding (in timber or plastics), typically for small-scale buildings, and lightweight cladding systems (such as curtain walling, rainscreen overcladding), typically for large-scale ones. The scale of the defects may also be influential. References on lightweight cladding systems for overcladding are limited and those given below provide basic guidance only.

For slating and tiling:

(a) BS 5534: Part 1: 1978: *Code of Practice for slating and tiling* (formerly CP 142: Part 2).

(b) Handisyde, C. C., *Everyday details* 9, 'Tile hanging and timber cladding'.

For lightweight cladding systems:

(c) BS 8200: 1985: *British Standard Code of Practice for design of non-loadbearing external vertical enclosures for buildings*, deals with functional requirements and principles (no details).

(d) BRE Report, *Overcladding external walls of large panel systems*, by H. W. Harrison, J. H. Hunt and J. Thomson, addresses all the main issues and has case

studies. Principles apply to other constructions.

(e) 'Light cladding systems', by Barry Josey, *The Architect* (now returned as the *RIBA Journal*), September 1987, pages 61–81, is an excellent chatty review of what is involved and what is available.

(f) Brookes, Alan, *Cladding of buildings*, is a good review of a wide range of claddings with details.

(g) Brookes, A. J. and Grech, C. *The building envelope*, Applications of new technology cladding, Butterworth Architecture, 1990. An excellent survey with 33 studies.

(h) Anderson, J. M. and Gill, J. R., *Rainscreen cladding, a guide to design principles and practice*, CIRIA/ Butterworths, 1988. Good for principles; more experience needed in practice.

6 For cavity fill:

(a) *BRE News* **51**, Spring/Summer 1980, 'Rain penetration tests on cavity-filled houses', by A. J. Newman and D. Whiteside, is a good summary of tests. Later tests on twelve cavity fills published in *Building and Environment*, **17**, No. 3, pages 175–191, 1982.

(b) BRE Digest 236, *Cavity insulation*, discusses forms of insulant available and recommends using blown-in rock fibre or polystyrene beads for least risk of rain penetration.

Note: The blown-in insulants have the added advantage that they can at least be taken out of the cavity if remedial work has to be carried out within the cavity to overcome problems of rain penetration.

Also draws attention to leakiness of walls having low-absorption bricks with recessed joints in outer leaf.

(c) *Cavity insulated walls*, Practice Note 1, jointly by ACBA (Aggregate Concrete Block Association Ltd), BDA, C&CA, Eurosil-UK, NCIA (National Cavity Insulation Association) and STA (Structural Insulation Association), September 1984. (Free on application to any of the organisations involved.) Broadsheet/do and do not format. Crisp and clear – good for checking on site. Specifiers guide below is more up-to-date and more informative.

(d) *Cavity insulated walls*, Specifiers guide, jointly by ACBA, British Plastics Federation (BPF), BDA, C&CA, Eurosil-UK, and NCIA, January 1987. (Free on application to any of the organisations involved.) Addresses the main issues concisely with clear, simple and crisp illustrations.

(e) BBA and others, *Cavity insulation of masonry walls – dampness risks and how to minimise them* (BBA, BRE, NFBTE and NHBC), November 1983. Comprehensive coverage of issues and details of the most common cavity fill in 'do and do not' data sheets.

(f) *Agrément Certificates* for individual products (at least 70 certificates by the end of 1987).

Note: Most, if not all, Agrément Certificates give the limitations of the product or the circumstances of its use. It is important that these are read and interpreted for particular cases. See also Information sheets below.

(g) BBA, Information sheet 10, *Methods of assessing the exposure of buildings for cavity fill insulation* (revised edition, January 1983), relates to and formed the basis of BS 5618.

(h) BBA, Information sheet 16, *Cavity wall insulation*, has important qualifications as to the suitability of certain walls for cavity fill, including the interpretation of local practice and for severe exposure conditions as well as the increased risk of water penetration when low-porosity bricks and/or recessed joints are used. These qualifications were seldom in certificates for particular products.

(i) BS 5618: 1985: *Code of Practice for the thermal insulation of cavity walls (with masonry or concrete inner and outer leaves) by filling with urea–formaldehyde foam systems.* This is a new edition of the code that updates suitable properties and tightens the installation procedure and supersedes the 1978 edition.

(j) BS 8280: Part 1: 1985: *Guide to the assessment of suitability of external cavity walls for filling with thermal insulants,* 'Existing traditional cavity construction'. Guidance on cavity walls not exceeding 12 m high or cavity walls without vertical members of the structural frame bridging the cavity. No guidance is given on design and construction of new work.

(k) BS 6676: Part 2: 1986: *Code of Practice for installation of batts (slabs) filling the cavity construction,* gives criteria for design and construction of walls to be insulated; stresses precautions. Appendix gives additional recommendations for walls over 12 m high.

(l) Fibreglass Ltd, *Cavity wall test facilities,* by Rick Wilberforce, Building Technical File, 3, October 1983, pages 51–53, is useful as background to testing. Read with 'Cavity fill' below.

(m) Fibreglass Ltd, *Feedback on cavity fill,* by L. W. Jackson, Building Technical File, 6, July 1983, pages 53–58, is a useful reminder as to the need for all the precautions advised by Fibreglass for Dritherm Cavity Wall insulation that fills a cavity. The lessons apply to similar products by other makers.

(n) BRE DAS 17, *External masonry walls: insulated mineral fibre cavity width batts – resisting rain penetration,* is good on precautions.

(o) BRE DAS 79, *External masonry walls: partial cavity fill insulation – resisting rain penetration* (Site), on keeping cavities clean.

(p) BRE Digest 277, *Built-in cavity wall insulation for housing,* is good for points that need special consideration in design and construction.

(q) BRE GBG 5, *Choosing between cavity, internal and external wall insulation.* An excellent analysis of the alternatives when upgrading existing buildings.

Manifestation 3

More or less vertical cracking (sometimes vertical and diagonal) through the units and/or mortar joints, with or without dampness on the inside surface. The cracking often occurs in the 'sprandrel' between openings.

Cause

1 Inadequate provision of movement joints for moisture movement in masonry.
2 If there are many fine cracks and water penetration, the damp is probably due to other defects such as faulty dpcs and cavity trays – see Study 16: Window openings.

Cross reference

Study 24: Movements under heading 'Movement joints', 8.05, page 144.

Remedial work

1 *At best* if the cracking is confined to the mortar joints:
  (a) Rake out and repoint with the weakest possible mortar.
2 *At worst* if the units themselves are cracked:
  (a) Take out the cracked units only and rebuild with a movement joint; or in very severe cases:

(b) Take down part or all of the outer leaf and rebuild incorporating a movement joint.

References for correct detailing

1 BRE Digest 157, *Calcium silicate (sand lime, flint lime) brickwork,* describes details of movement joints on page 5. These should be provided at intervals of 7.5–9 m.
2 BS 5628: Part 3: 1985: *British Standard Code of Practice for use of masonry,* 'Materials and components, design and workmanship' (formerly CP 121: Part 1: 1973). Spacing of movement joints at intervals of between 7.5 and 9 m – as in BRE Digest 157, clause 20.3.2.3, page 33. Mortar mixes in clause 23, page 63, and Table 15 of the same page.
3 BRE Digest 362, *Building mortar.* Gives recommendations for the composition and use of general purpose mortar and other specialised types of mortar. The recommendations reflect changes in British Standards and impending changes from British to European standards.
4 BRAS, *Cracking in calcium silicate brickwork,* Building Technical File, 8, January 1985, pages 61–63, is a report of an investigation on cracking in two housing estates. Useful summary of the main issues, including repair.
5 Eldridge, H. J., *Common defects in building.* The problems are discussed on pages 103–105.

Manifestation 4

Horizontal cracking in mortar joints in the outer leaf, at about 300 mm intervals or every fourth course, with or without dampness on the inside surface. Exceptionally, some horizontal cracking of plaster on the inner leaf. This type of cracking can be distinguished from sulphate attack by the predominance of irregular horizontal cracks.

Cause

Corrosion of ferrous metal ties in black ash mortar due to the conversion of sulphides in the ash to sulphates. Any chlorides in the ash will aggravate the corrosion. Galvanising of the ties offers no long-term protection.

Cross reference

Study 26: Corrosion. See 'Corrosion in cavity wall ties', 6, page 160.

Remedial work

1 If a long-term remedy is required and corrosion of the wall ties is advanced, rebuild the outer leaf using stainless steel ties. Alternatively, and more economically, replace the corroded ties individually. It is now possible to replace wall ties using resin grouted rods. Neither cladding nor rendering is recommended as a long-term remedy.
2 If a long-term remedy (25 years at least) is required, the ties are not badly corroded and the building is a dwelling of no more than two storeys, then the cavity wall can be stabilised by filling the cavity with an *in-situ* heavy-duty polyurethane foam, and installed by specialist contractors.
3 If a short-term remedy is required and the ties are not badly corroded (or only some ties have been affected) replace or treat the affected ends or apply wall cladding to reduce the rate of further corrosion.

References for correct detailing

1 BRE Information Paper IP 28/79, *Corrosion of steel wall ties,* deals fully with occurrence, diagnosis and appropriate remedial action.

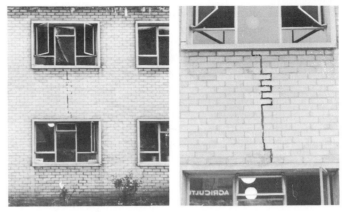

14 *Vertical cracks extending through the walling units themselves (concrete blocks).*

16a *Cracking along joints, made good with mastic (sand lime bricks).*
16b *Close-up of typical patterns.*

2 BRE Information Paper IP 29/79, *Replacement of cavity wall ties using resin grouted stainless steel rods*. This method is reasonably inexpensive and easy to carry out on site.

Note: These two Information Papers supersede: BRAS TIL 22, *Corrosion of wall ties in cavity brickwork.*

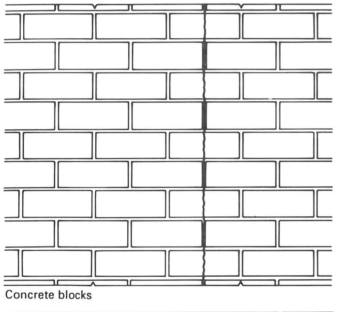

Concrete blocks

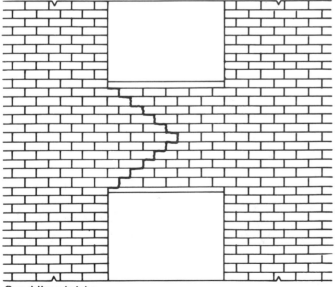

Sand-lime bricks

15a, b *Two alternative types of failure.*

3 Eldridge, H. J., *Common defects in building*, pages 125–126 for remedies but excluding resin bonded techniques. Updated in PSA *Defects in building*, 5.1.12, pages 74, 75 illustrating repair methods.
4 BRE Current Paper CP 3/81, *The performance of cavity wall ties.*
5 BRE Information Paper 4/81, *The performance of cavity wall ties*, a précis of the CP above, summarises the results of a recent survey of galvanised steel wall ties in high- and low-rise buildings of various ages. The rate of corrosion was found to be greater than has been assumed hitherto, with potentially serious implications for exposed high-rise and poorly braced cross-wall buildings. The paper recommends that from now on alternative ties are used. Until plastic ties have been more fully tested, stainless steel ties are to be preferred.
6 BRE Digest 329, *Installation of wall ties in existing construction*, deals comprehensively with the problem. The earlier IP 28/79 is better for diagnosis.
7 BRE DAS 19, *External masonry cavity walls: wall ties – selection and specification*, reminds on need for proper type and frequency of installation.
8 BRE DAS 20, *External masonry cavity walls: wall ties – installation*, is site application of advice in DAS 19.
9 BRE DAS 21, *External masonry cavity walls: wall tie replacement*, describes diagnosis and remedial work.
10 BRE Information paper, IP 6/86, *The spacing of wall ties in cavity walls*. Test results confirmed type of tie required and recommendations in BS 5628 as to spacing.

Manifestation 5
Shelling and bulging of finishing plastercoat internally (may also occur on internal walls or partitions).

Cause
Weakness in the bond between the plaster and its background, or between successive layers of plaster; and/or weak bond broken by shear stresses resulting from differential movement between the plaster and its background (drying shrinkage of the background is a notable cause).

Cross reference
Study 25: Loss of adhesion, under heading '5 Gypsum plaster finishing coats on lightweight concrete blockwork', page 150.

References for correct detailing
1 Building Research Advisory Service TIL 14, *Shelling of plaster finishing coats*. Undercoats must be allowed to dry

sufficiently to develop adequate suction and thereby a good bond with the finishing coats. Under tests, finishing coats applied seven days after the undercoat resisted three times the compressive strain as those applied after twenty-four hours. For remedial work, a PVAC bonding solution is recommended.

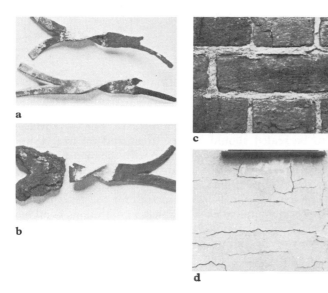

17a, b *Corroded wall ties. The part of the tie in the outer leaf is usually far more severely corroded than that in cavity or inner leaf. Expansion of corroded part then causes crack in horizontal joint in outer leaf of wall, in which it is embedded.*
17c, d *Horizontal cracks of kind often associated with corrosion of wall ties, in brickwork joints and in rendering. (Building Research Establishment, Crown Copyright).*

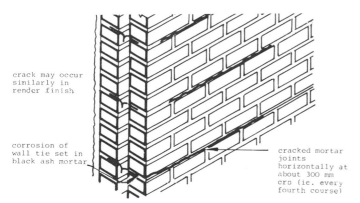

crack may occur similarly in render finish

corrosion of wall tie set in black ash mortar

cracked mortar joints horizontally at about 300 mm crs (ie. every fourth course)

18 *Physical condition associated with Manifestation 4.*

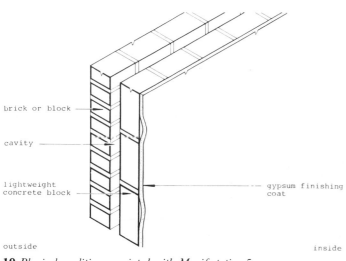

brick or block

cavity

lightweight concrete block

gypsum finishing coat

outside   inside

19 *Physical condition associated with Manifestation 5.*

2 BS 5492: 1977: *Code of Practice for internal plastering*, gives complete and thorough guidance.
3 BRE Digest 213, *Choosing specifications for plastering*, is organised as a sequence of decisions to be taken, effectively summarising BS 5492 (reference 2 above).
4 BRAS, *Cracking and bulging of plaster*, Building Technical File, 4, January 1984, pages 45 and 46, uses investigation of plaster defects in one house to highlight problems with plastering and includes note on remedial measures.
5 Eldridge, H. J., *Common defects in buildings*. The problem is discussed in detail on pages 227–228. Updated in PSA *Defects in buildings*, 5.3.14, pages 190, 191.

Manifestation 6
White deposit on the surface of brickwork, often mistaken for efflorescence. It can be distinguished from efflorescence in that the deposit is not removed either by brushing or by rain.

Cause
1 Free lime leached from cement- and lime-based products (notably mortar and its droppings in cavities) within wall constructions, by water transferred through the product, is brought to the exposed surface where it is converted to an insoluble form of lime.
2 Leaking through mortar joints (low absorption bricks notable); saturation of the walling because of inadequate damp-proofing and/or lack of adequate protection during construction.

Remedial work
1 At best, if the lime staining is the result of inadequate protection during construction: Remove deposit with a chemical treatment after testing a trial area.
2 At worst, if the lime staining is the result of poor jointing or inadequate damp-proofing:
   (a) Remove deposit chemically with care. Rake out all joints to a depth of at least 15 mm and repoint.
   (b) Remove deposit chemically with care and reinstate faulty damp-proofing/brickwork or clad relevant parts of the wall.

References for correct detailing
Addleson, Lyall and Rice, Colin, *Performance of materials in buildings*, Butterworth-Heinemann, 1992, 3.5 Efflorescence. 'Lime staining', pages 371–372.

20 *Two examples of lime staining (Manifestation 6).*
a *New brickwork soon after completion. Lime has found its way through weepholes.*
b *Intense hard staining, confined mostly to the surface of (hard) bricks. The path of the lime was through the softer mortar.*

Study 14
RC frame/cavity wall junctions

E 3

Manifestation 1
Detachment and/or degradation of the sealant in movement or expansion joints, with or without rain penetration which may cause dampness inside the building. It can occur in a wide variety of constructional details.

Cause
1 If the sealant is detached:
   (a) Inability of the sealant to accommodate movement and/or
   (b) Inadequate preparation of all surfaces accepting the sealant.
2 If the sealant is degraded:
   (a) Excessive exposure to sunlight (i.e. there has been inadequate protection of the sealant) or
   (b) Vandalism (either by humans or birds) of soft material.

Cross references
Study 24: Movements under heading 'Movement joints' in particular, 8.05, page 144.
Study 25: Loss of adhesion under heading '8 Sealants at joints, walls and around windows', page 152.

Remedial work
Remove all defective material, thoroughly clean all surfaces to receive the new sealant – a difficult but essential job.

Then apply correctly an appropriate sealant in accordance with the manufacturer's instructions, suitably protected against sunlight. Important factors to consider include:
(a) Adequate preparation of the surfaces. The two surfaces to be sealed should be dry, dust and oil free. Prime where necessary in accordance with the manufacturer's instructions (not all sealants require priming of the surfaces of the joint).
(b) The joint should be wide enough for the sealant to be physically applied ensuring that a continuous bond has been made. It should be wide enough for a sealant with a given flexibility to accommodate the estimated movement of the joint – this may mean extra chasing.
(c) Use of a back-up strip and/or bond-breaking tape. Sealants are designed to be bonded to two surfaces only

2 *Loss of adhesion and cracking in sealant. Note also crack in parapet beam over joint in brickwork, where no movement joint has been provided.*

1 *Designers often have unrealistic expectations of mastic performance. The mastic sealant in joint to parapet below coping has suffered loss of adhesion and detachment; opening up between the mastic and the sides of the joint indicates the amount of movement.*

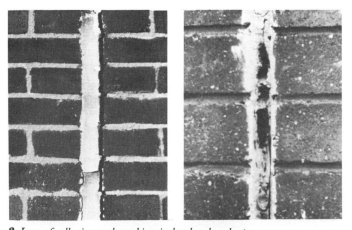
3 *Loss of adhesion and cracking in hardened sealant.*
4 *Degradation of sealant, aggravated by vandalism (dark areas indicate where mastic has been picked out of joint).*

so that their movement is unrestrained except at those two surfaces. A resilient foam back-up strip ensures that the sealant does not spread into the depth of the joint but is forced against the two surfaces being sealed.
(d) Where fillets are used for pointing they should have a convex profile.

References for correct detailing
1 Sealant Manufacturer's Conference and CIRIA: *Manual of good practice in sealant application*, sets out the principles of both selection and application. The illustrations are not good and it is due for an update.
2 BRE Information paper IP 25/81, *The selection and performance of sealants* – see update below.
3 BRE, *Selection, performance and replacement of building joint sealants*, by J. C. Beech, *Building Technology and Management*, 1983, Part 1, 21(7) pages 23–25 and Part 2, 21(8) pages 14–16.
4 BRE, *Failures with site-applied adhesives* by J. R. Coad and D. Rosaman, Building Technical File, 7, October 1983, pages 57–62. Not related specifically to sealants but useful as to precautions with materials intended to adhere to others – see update below.
5 BRE Information Paper IP 12/86, *Site-applied adhesives – failures and how to avoid them*. Update of Building Technical File article – see above.
6 BS 6093: 1981: *Code of practice for design of joints and jointing in building construction*, gives good coverage of functional requirements and joint types.

7 Martin Bruce, *Joints in building*, has wide coverage of principles and practice and complements code above.
8 BS 3712: 1987: *Building and construction sealants*, has various parts dealing with methods of test for different properties of sealants.
9 BS 4254: 1983: *Specification of two-part polysulphide based sealants*.
10 *AJ* Series, *Products in Practice* and recently *AJ Focus*. For review of products available. No specific issues noted as coverage changes fairly frequently.

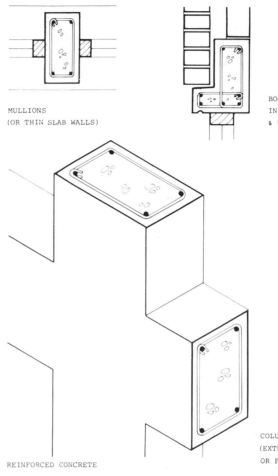

MULLIONS
(OR THIN SLAB WALLS)

BOOT LINTELS
IN FRAMED
& UNFRAMED WALLS

REINFORCED CONCRETE

COLUMNS & BEAMS
(EXTERNAL) CLADDING
OR FACING OMITTED

5, 6 *Example of cracking and dislodgement of mosaic finish as a result of corrosion and resultant expansion of reinforcement (photo GLC); and cracked edge of supporting slab, exposing reinforcement.*
7 *Cracking and rust-staining of thin concrete mullion.*
8 *Typical positions in the vicinity of which the above manifestation may occur.*
9 *Surface cracking to exposed concrete column, betraying corrosion of reinforcement beneath surface. (Photos 6 and 9 Building Research Establishment, Crown Copyright)*

Manifestation 2

Cracking of concrete and/or the rusting of adjacent areas.

Condition

The schematic diagram (**9**) shows constructional details in which the manifestation of failure may have arisen. It is a generalised drawing and the principles illustrated must be interpreted in relation to the specific construction found on site in any particular case.

Cause

Corrosion of mild steel reinforcement due to one or more of the following:

1 Inadequate concrete covers to the mild steel for the grade and type of concrete used and its degree of exposure.
2 Addition of excess amounts of calcium chloride to accelerate setting.
3 The use of aggregates containing chlorides, not necessarily calcium chloride.
4 Less commonly, carbonation of the concrete.

Cross reference

Study 26: Corrosion under heading '3 Mild steel reinforcement in concrete', page 157.

Remedial work

1 *At best*:
   If the cracks are fine and there has been some serious dislodgement of the concrete, scrape out the cracks and fill them with a cement:sand mix with a bonding/waterproofing agent such as Ronafix.
2 *At worst*:
   (a) Repair: firms who are experienced in repairing damaged concrete with epoxy resins, gun spraying and other specialised techniques should be consulted and used.
   (b) Overcladding but only after badly corroded steel reinforcement has been properly protected.

References for correct detailing

1 For concrete mixes:
   (a) BS 8110: Part 1: 1985: *Structural use of concrete: Code of Practice for design and construction*, includes guidance on specification and workmanship. (Replaces CP 110: Part 1: 1972.)
   (b) BRE Digest 237, *Materials for concrete*, is good on mix design/functional requirements.
   (c) BRE Digest 244, *Concrete mixes: specification, design and quality control*, relates to Digest 237.
   (d) BRE, *Design of normal concrete mixes*, is standard guide on concrete mix design and now updated to reflect changes in materials.
2 For corrosion/protection:
   (a) BRE Digest 59, *Protection against corrosion of reinforcing steel in concrete*.
   (b) BRE Information paper IP 12/80, *Deterioration due to corrosion in reinforced concrete*, includes diagnosis and repair.
   (c) BRE Digest 109, *Zinc-coated reinforcement for concrete*.
   (d) BRE Digest 263, *The durability of steel in concrete: Part 1 Mechanism of protection and corrosion*.
   (e) BRE Digest 264, *The durability of steel in concrete: Part 2 Diagnosis and assessment of corrosion-cracked concrete*.
   (f) BRE Digest 265, *The durability of steel in concrete: Part 3 The repair of reinforced concrete*.

3 For cracking/diagnosis:
   BRE Reports on investigations into the condition of a number of post-war prefabricated reinforced concrete houses/systems. These are not listed here but could usefully be consulted for problems associated with corrosion of the steel reinforcement. Some have excellent coloured photographs.
4 For carbonation:
   BRE Report 75, *Depths of carbonation in structural-quality concrete: an assessment of evidence from investigations of structures and from other sources*.
5 For overcladding:
   (a) BS 8200: 1985: *British Standard Code of Practice for design of non-loadbearing external vertical enclosures for buildings*, deals with functional requirements and principles. No details.
   (b) BRE Report, *Overcladding external walls of large panel systems*, by H. W. Harrison, J. H. Hunt and J. Thomson, addresses all the main issues and has case studies. The principles are equally applicable to other forms of construction.
   (c) 'Light cladding systems', by Barry Josey, *The Architect* (now returned as the *RIBA Journal*), September 1987, pages 61–81, is an excellent chatty review of what is involved and what is available.
   (d) Brookes, Alan, *Cladding of buildings*, is a good review of a wide range of claddings with details.
   (e) Anderson, J. M. and Gill, J. R., *Rainscreen cladding, a guide to design principles and practice*, CIRIA/Butterworths, 1988. Good for principles; more experience needed in practice.

Manifestation 3

Internal dampness at or near floor level occurring at or near columns. Finishes such as carpets are usually affected.

Caution: The location in which dampness may manifest itself is not indicative of the source of entry of the water. All the causes listed under Manifestations 1, 2, 4 and 5 should therefore be given serious consideration, even in seemingly obvious cases. Further 'clues' may show up during remedial work, and modify the diagnosis.

Condition

The schematic diagram (**10**) shows the constructional detail in which the above manifestation of failure has arisen.

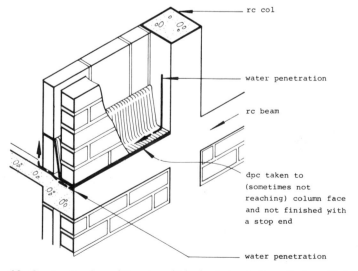

rc col

water penetration

rc beam

dpc taken to (sometimes not reaching) column face and not finished with a stop end

water penetration

10 *Constructional condition in which damp penetration of above kind commonly occurs in reinforced concrete/brick panel construction.*

It is a generalised detail and the principles illustrated must be interpreted in relation to the specific construction found on site in any particular case.

Cause
1 If damp is associated with rain:
   The water finds its way past the outer skin (which is acceptable) but then penetrates around the junction of a dpc and the sides of the column. It then enters at the floor line. The amount of water penetration is likely to be increased if the wall is severely exposed to driving rain; and/or if there is no drainage of the cavity by means of open perpends, weepholes, etc.
2 If damp is *not* associated with rain:
   If the incidence of damp manifestation cannot be correlated with periods of heavy or continuous rain (particularly rain associated with driving wind) but rather with cold weather, then the cause is likely to be cold bridging, resulting in surface condensation. In this connection see also Study 19: Floors and ceilings.

Cross references
Study 22: Rain penetration under heading 'Cavity walls: dpcs and cavity trays', 8.04, page 127.
Study 20: Condensation under heading 'Surface condensation: Cold bridges', 4.02, page 108.

Remedial work
1 If associated with rain:
   Take out sufficient of the outer leaf at the junction between the wall panel and the rc column to allow for the insertion of a stop end dp tray, either sealed to the column or counter-flashed, to prevent water running down the column from penetrating behind the dp upstand. There should be a 100 mm lap between the dpc and the new stop end.
2 If not associated with rain:
   Increase the thermal insulation of the wall/floor junction by applying an adequate thickness of extruded polystyrene, at least 450 mm along the floor and 300 mm up the wall. (Extruded polystyrene has the advantage over other insulants in having the properties of a vapour check. If other insulants are proposed which are vapour permeable, such as expanded polystyrene, a vapour check on their warm side will be required.) The thermal insulation has to be applied in a vulnerable position and will therefore need to be adequately protected. Such protection is difficult to achieve in practice. In the floor it may be possible to cut out a 'chase' in the screed to accept the thermal insulant; on the wall some form of protective lining and/or skirting has to be devised.

References for correct detailing
1 Duell and Lawson, *Damp-proof course detailing*, Fig. 39, page 20.
2 DOE, *Construction*, **14**, pages 18 and 19, Figs. 10 and 12.

Manifestation 1
Internal dampness at or near floor level occurring at or near columns. Finishes such as carpets are usually affected.

Caution: The location in which dampness may manifest itself is not indicative of the source of entry of the water. All the causes listed under Manifestations 1, 2, 3 and 5 should therefore be given serious consideration, even in seemingly obvious cases. Further 'clues' may show up during remedial work, and modify the diagnosis.

Condition
The schematic diagram **11** shows the constructional detail in which the above manifestation of failure has arisen. It is a generalised detail and the principles illustrated must be interpreted in relation to the specific construction found on site in any particular case.

Cause
Water is finding its way underneath the dpc. Note that penetration may be increased for the reasons given under Manifestation 3, and in addition if the adhesion between the dpc and the concrete is poor – some plastics dpcs are reported not to have good adhesion.

Cross reference
Study 22: Rain penetration under heading '8 Cavity walls: dpcs and cavity trays', 8.04, page 127.

Remedial work
The existing dpc must be extended so that rainwater is thrown clear of the face of the building. Although complicated to carry out, the only effective way of doing this is by cutting out the course of bricks above the dpc and then inserting a non-ferrous metal tray under the existing dpc. The new metal tray (sometimes known as a tingle) should have its back edge folded over and its front edge projecting over the concrete beam. In executing the work, it is necessary to have short sections of the bricks above the dpc (about three or four bricks) removed at any one time. Care should be taken not to damage the existing dpc.

References for correct detailing
1 DOE, *Construction*, **14**, page 18, Fig. 10.
2 Duell and Lawson, *Damp-proof course detailing*, Fig. 29, page 16.

Both the references show the correct details in new work.

Manifestation 5
Internal dampness at or near floor level occurring at or near columns. Finishes such as carpets are usually affected.

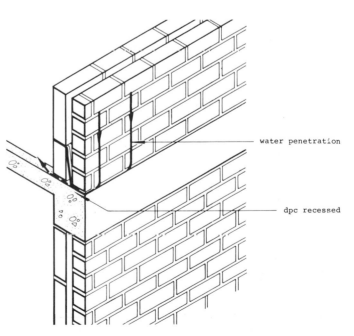

11 *Constructional condition with which water penetration at floor line, but not localised at columns, is frequently associated.*

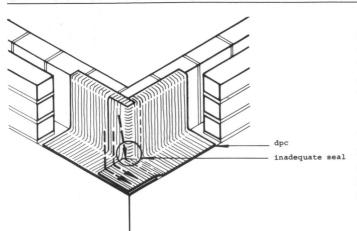

EXTERNAL CORNER

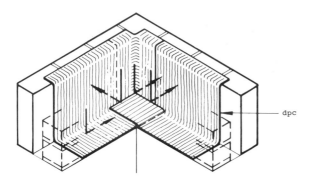

INTERNAL CORNER

12 *Two dpc conditions allowing water penetration at or near corners — sealing at changes of direction is often negelcted.*

Caution: The location in which dampness may manifest itself is not indicative of the source of entry of the water. All the causes listed under Manifestations 1, 2, 3 and 4 should therefore be given serious consideration, even in seemingly obvious cases. Further 'clues' may show up during remedial work, and modify the diagnosis.

Condition
The schematic diagram, **12**, shows the constructional detail in which the above manifestation of failure has arisen. It is a generalised detail and the principles illustrated must be interpreted in relation to the specific construction found on site in any particular case.

Cause
Water is finding its way past the dp tray at the corners and at laps due to inadequate sealing of the laps. Note that penetration may be increased if the wall is severely exposed to driving rain; and/or if there is no drainage of the cavity by means of open perpends, weepholes, etc. It may be substantially worse if the vertical dpc is sloped and worse still if the sloped dpc is unsupported: a sagging dpc tends to form gaps at the laps.

Cross reference
Study 22: Rain penetration under heading '8 Cavity walls: dpcs and cavity trays', 8.04, page 127.

References for correct detailing
1 DOE, *Construction*, **14**, page 19, Fig. 12.
2 Duell and Lawson, *Damp-proof course detailing*, Figs. 41,a,b, page 21 and Figs. 57–61, pages 29 and 30. This reference gives very clear drawings of how to form waterproof dpcs at internal and external corners. Some manufacturers make preformed dpc corner pieces. (Calenders, the makers of lead core, bitumen and felt dpcs, claim that their material can be formed in much the same way as pitch polymer dpcs. Consult the manufacturer for his recommendations.)

Study 15
RC frame/concrete panel junctions

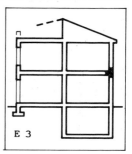

E 3

Manifestation 1
Dampness internally, at or near panel joints.

Condition
The schematic diagram **2** shows the constructional detail in which the above manifestation of failure has arisen. It is a generalised detail and the principles illustrated must be interpreted in relation to the specific construction found on site in any particular case.

Cause
Rain penetration, especially if the walls are subject to driving rain.
1 Defective flashing:
  (a) If the flashing edge is recessed behind the outer face of the concrete panel, rainwater penetrates underneath the flashing and there is no protection or alternative line of defence behind (see section 1 in **3**) to prevent further water penetration.
  (b) If the flashing has been omitted (see section 2 in **3**) protection behind the baffle is greatly reduced.
2 Defective baffle:
  Rainwater penetrates behind the baffle.

Cross reference
Study 22: Rain penetration under heading 'Joints in concrete panels,' 10, page 130.

References for correct detailing
1 For open drain joints:
  (a) Martin, B., *Joints in building*, figures on pages 115–117.
  (b) BRE Digest 85, *Joints between concrete wall panels: open drain joints*, Fig. 5, page 3.
  (c) DOE, *Construction* **13**, page 15, Fig. 2.
2 For sealed joints:
  (a) Sealant Manufacturer's Conference and CIRIA: *Manual of good practice in sealant application* sets out the principles of both selection and application. The

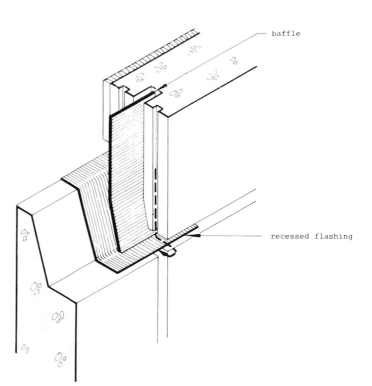

2 *One common condition of failure: flashing is recessed behind panel face and rainwater penetrates beneath it, especially at vertical joint.*

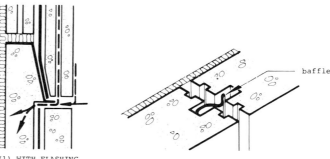

(1) WITH FLASHING
SECTIONS

baffle

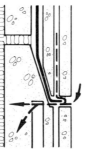
(2) WITHOUT FLASHING

3 *Above left, section showing inadequate flashing; below left, no flashing at all. Above, baffle is often deformed or otherwise defective, and does not seal at jambs of recessed joint.*

1 *Typical baffle-jointed concrete panels: joints are open, not sealed, and rely on vertical baffles and horizontal flashings to exclude rainwater.*

illustrations are not good and it is due for an update.

(b) BRE Information paper IP25/81, *The selection and performance of sealants* – see update below.

(c) BRE, *Selection, performance and replacement of building joint sealants,* by J. C. Beech, *Building Technology and Management*, 1983, Part 1, 21(7) pages 23–25 and Part 2, 21(8) pages 14–16.

(d) BS 6093: 1981: *Code of Practice for design of joints and jointing in building construction* has good coverage of functional requirements and joint types.

(e) Martin Bruce, *Joints in building* gives a wide coverage of principles and practice and complements Code above.

(f) BRE, *Failures with site-applied adhesives* by J. R. Coad and D. Rosaman, Building Technical File, 7, October 1983, pages 57–62, is not related specifically to sealants but useful as to precautions with materials intended to adhere to others – see update below.

(g) BRE Information Paper IP12/86, *Site-applied adhesives – failures and how to avoid them* is an update of Building Technical File article – see above.

(h) BS 3712: 1987: *Building and construction sealants* has various parts dealing with methods of test for different properties of sealants.

(i) BS 4254: 1983: *Specification of two-part polysulphide based sealants*.

(j) *AJ* Series *products in Practice* and recently *AJ Focus*. For review of products available. No specific issues noted as coverage changes fairly frequently.

3 General

BS 6093: 1981: *Code of practice for the design of joints and jointing in building construction* comprehensively deals with the problem of joints and jointing.

Study 16
Window openings

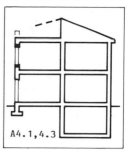

A4.1,4.3

Manifestation 1
Dampness internally around the opening or in the reveals to the opening, sometimes with mould growth.

When considering 'Causes' and 'Remedial work' under each Condition, note that filling the cavity has the effect of highlighting or revealing faults that in an unfilled cavity may not cause problems.

Remedial work can present special difficulties (Study 13, Manifestation 1, pages 66–68).

Condition 1
The schematic diagram, **2**, shows the constructional details in which the above manifestation of failure has arisen. It is a generalised detail and the principles illustrated must be interpreted in relation to the specific construction found on site in any particular case.

Cause 1: associated with rain
Water penetrating the outer leaf and finding its way to the inner leaf at one or more of the encircled positions shown in **2**:

1 The water can penetrate at the joint and then run along the underside of the dpc and overflow at the end of the lintel (**2a** and **b**).

   Note: Greater risk of penetration if the dpc is unsupported on the slope: sagging of the dpc causes a gap to open between the two lapped layers.

2 Water running down the inner face of the outer leaf can overflow at the ends of the lintel and then penetrate to the inner leaf.

   Note: The risk of penetration is increased by the presence of a ledge formed by the end of the lintel not overlapping the cavity closing at the jamb.

3 Similar penetration to (1) above if the dpc is damaged during cleaning out of the cavity, especially if the dpc is unsupported on the slope (**2a** and **b**).

1 Clumsily built dp tray discharging water into the cavity.

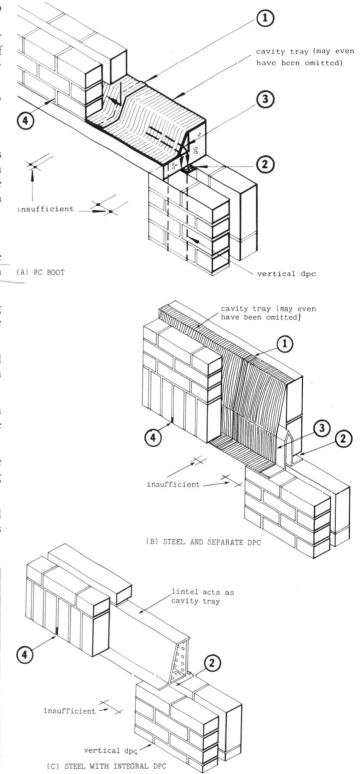

(A) RC BOOT

cavity tray (may even have been omitted)

insufficient

vertical dpc

cavity tray (may even have been omitted)

insufficient

(B) STEEL AND SEPARATE DPC

lintel acts as cavity tray

insufficient

vertical dpc

(C) STEEL WITH INTEGRAL DPC

2 Circled numerals refer to text (where numbers given refer to possible causes of failure).

4 Absence of open vertical joints for drainage over the lintel (long lintels particularly increase the risk of penetration if water dams on the top of the dpc). Alternatively, inappropriately designed drainage where exposure to driving rain is severe may allow water to penetrate the outer leaf more readily. In such cases L-shaped weep tubes pointing down or V-shaped venturis are better than open perpends to prevent the blow back of rainwater into the cavity.

All the above causes are influenced by the effects of wind.

Cross reference
Study 22: Rain penetration, under heading 'Around windows: dpcs and cavity trays', 8.03, page 128.

Remedial work
It will almost certainly be necessary to open up from the outside – and in short lengths. A difficult and expensive process – more so if there is cavity fill.
Then:
1 Extend or replace the lintel in order to support a new piece of cavity tray which should extend 300 mm beyond the jamb;
2 Insert a new piece of cavity tray over the existing tray, providing a lap of at least 100 mm which must be sealed, or alternatively;
3 Replace any damaged dp tray and ensure that the sloping part is supported;
4 Provide open vertical joints (weepholes) every fourth brick, where drainage of the cavity is inadequate. In very exposed areas, L-shaped weepholes or T-shaped venturis will be required (see references below).

Notes:
See Study 13, Manifestation 2.

1 For potential leakiness of soldier course brickwork – Condition 1, Note to 1(c), page 68.
2 For use of cladding that is part of comprehensive remedial work – References for correct detailing, p. 97.

References for correct detailing
1 Handisyde, C. C., *Everyday details* 10, 'Cavity wall lintels', deals comprehensively with different lintel types.
2 Duell and Lawson, *Damp-proof course detailing*, Fig. 19c,d, page 15 and Fig. 52b, page 26, for correct lintel dpc details and Fig. 35, page 17, for weephole details.
3 BRE DAS 12, *Cavity trays in external walls: preventing water penetration* (Design). Includes lintels.
4 BS 5628: Part 3: *British Standard Code of Practice for use of masonry*, 'Materials and components, design and workmanship' (formerly CP 121: Part 1: 1973). Fig. 12, pages 48 and 49.
5 *Cavity insulated walls*, Specifiers guide, jointly by ACBA, British Plastics Federation (BPF), BDA, C&CA, Eurosil-UK, and NCIA, January 1987, page 9. Illustrations are simple and crisp.
BRE Report, *Thermal insulation: avoiding risks*, 2 Walls, 215, pages 16 for cold bridges and Fig 30, page 17 for other parts of the opening. Concise with excellent illustrations.

Cause 2: not associated with rain
If the incidence of damp manifestation cannot be correlated with periods of rain (particularly heavy or continuous rain), but rather with cold weather, then the cause is likely to be cold bridging, resulting in surface condensation.

Cross reference
Study 20: Condensation under heading 'Surface condensation: cold bridges', 4.02, page 108.

Remedial work
Increase the thermal insulation to the lintel by applying an adequate thickness of extruded polystyrene. The thermal insulant must also be taken at least 150 mm beyond the lintel. *Extruded* polystyrene has the advantage of having the properties of a vapour control layer. (If other insulants are proposed, which are vapour permeable, such as expanded polystyrene, a vapour control layer on their warm side will be required.) Composite panels of extruded polystyrene and plasterboard (Styroliner) and of expanded polystyrene, polythene and plasterboard (Gyproc) are now available. The thickness of the insulant will probably not be less than 15 mm (preferably a minimum of 25 mm). The minimum amount required can be found by calculation, following the methods described in the references listed at the end of Study 20: Condensation 5, page 112. Increased ventilation and higher levels of sustained heating may also be required.

Condition 2
The schematic diagram, **4**, shows the constructional detail in which the above manifestation of failure has arisen. It is a generalised detail and the principles illustrated must be interpreted in relation to the specific construction found.

Cause 1: associated with rain
Water is penetrating between the wall jamb and the window frame for one or more of the following reasons:

1 Loss of adhesion of the sealant pointing on diagram **1**. This can be due to the use of an incorrect sealant material; and/or the wrong application of the sealant for the amount of strain caused by differential movement between the window frame and the wall jamb. Note that moisture movement, especially of timber frames, can be significant.
2 In cavity walls, the vertical dpc is inadequate, **2**; and/or window frame wrongly positioned in relation to the dpc.

Cross references
Study 22: Rain penetration under heading 'Around windows: dpcs and cavity trays', 9.04, page 129.
Study 25: Loss of adhesion under heading 'Sealants at joints, walls and around windows', 8, page 152.

Remedial work
1 *At best*:
Take out the existing sealant and repoint with an appropriate sealant properly applied. Oleo-resinous and acrylic (emulsion) types of sealant are suitable, depending on the amount of movement and whether repainting will be carried out.
Important factors to consider include:
(a) Adequate preparation of the surfaces. The two surfaces to be sealed should be dry, dust and oil free. Prime where necessary in accordance with the manufacturer's instructions (not all sealants require priming of the surfaces of the joint).
(b) The joint should be wide enough for the sealant to be physically applied ensuring that a continuous bond has been made. It should be wide enough for a sealant with a given flexibility to accommodate the estimated movement of the joint. This may mean extra chasing.
(c) Use of a back-up strip and/or bond-breaking tape. Sealants are designed to be bonded to two surfaces

only so that their movement is unrestrained except at those two surfaces. A resilient foam back-up strip ensures that the sealant does not spread into the depth of the joint but is forced against the two surfaces being sealed.

(d) Where fillets are used for pointing they should have a convex profile.

2 *At worst*:

Remove and replace the vertical dpc. This is an extremely difficult operation and probably only feasible if the dpcs elsewhere around the opening are also faulty. Proper pointing with a sealant may provide a cure, subject to the degree of exposure to driving rain and should be tried first. The sealing may be improved if, before the sealant is applied, the joint is partially filled with a polyurethane foam, such as Bostik 995.

Cause 2: not associated with rain
Surface condensation (usually betrayed by the presence of mould growth) due to cold bridging. See 3 on the diagram.

Cross reference
Study 20: Condensation under heading 'Surface condensation: cold bridges', 4.02, page 108.

Remedial work
Increase the thermal insulation of the jambs by applying an adequate thickness of *extruded* polystyrene. This has the advantages over other insulants in having the properties of a vapour control layer. If other insulants are proposed which are vapour permeable, such as expanded polystyrene, a vapour control layer on their warm side will be required. Both extruded polystyrene and expanded polystyrene are available as composite panels lined with plasterboard (and a polythene film if expanded polystyrene is used), as Styroliner and Gyproc, respectively. It is most important that the insulant is fully bonded to the background to avoid the risk of interstitial condensation occurring at the interface of the insulant and the background. The thickness of the insulant will probably not be less than 15 mm. The minimum amount required can be calculated by following the methods described in the references at the end of Study 20: Condensation 5, page 112. Increased ventilation and/or higher levels of sustained heating may be necessary.

References for correct detailing
1 For the selection of sealants for pointing:
(a) BRE DAS 68, *External walls: Joints with windows and doors – detailing of sealants.* (Design) has a list of sealants suitable for use around windows.
(b) BRE DAS 69, *External walls: Joints with windows and doors – application of sealants.* Diagrams illustrate the essential points when sealants are applied.
(c) Sealant Manufacturers' Conference and CIRIA: *Manual of good practice in sealant application.* Sets out the principles both of selection and application – poor illustrations.
2 For correct vertical dpc details:
(a) Duell and Lawson, *Damp-proof course detailing*, Fig. 19, page 13 and Fig. 51, page 27.
(b) BS 5628: Part 3: 1985: *British Standard Code of Practice for use of masonry*, 'Materials and components, design and workmanship' (formerly CP 121: Part 1: 1973). Fig. 12, pages 48 and 49.
(c) BRE Report, *Thermal insulation: avoiding risks*, 2 Walls, 215, pages 16 for cold bridges and Fig 30, page 17 for

3 *Example of loss adhesion of mastic, leading to water penetration.*

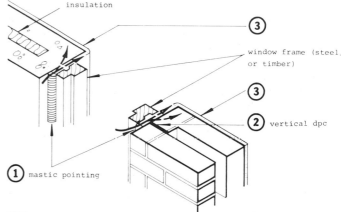

4 *Circled numerals refer to numbers given in parentheses in text.*

other parts of the opening. Concise with excellent illustrations.

Condition 3
The schematic diagram **5** shows the constructional detail in which the above manifestation of failure has arisen. It is a generalised detail and the principles illustrated must be interpreted in relation to the specific construction found on site in any particular case.

Cause 1: only associated with rain
Water penetrating at weaknesses in the dpc for one of the following reasons:

1 The dpc may have been omitted altogether (**1**).
2 The laps between the lengths of dpc are inadequate (**2**) in which case water has gained access at joints between brick, tile or concrete sill units (**3**) which have opened up due to movement (i.e. there has been a loss of adhesion of the mortar joints).
3 An inadequately formed junction (**4**) between the horizontal (sill) dpc and the vertical (jamb) dpc. The water penetration may be aggravated by the joint opening up (cracking and/or loss of adhesion of the mortar joints, or loss of adhesion of the sealant pointing) due to inadequate provision for differential movement between the sill and the jamb (**5**).

Cross references
Study 22: Rain penetration under heading 'Cavity walls, dpcs and cavity trays', 8.04, page 127.
Study 25: Loss of adhesion under headings 'Mortars', 4, page 147, and '8 Sealants at joints, walls and around windows', page 152.

Remedial work
1 *At best*:
Fill the joints that have opened up at points (**3**) and (**5**) with an appropriate sealant.
Important factors to consider include:
(a) Adequate preparation of the surfaces. The two surfaces to be sealed should be dry, dust and oil free. Prime where necessary in accordance with the manufacturer's instructions (not all sealants require priming of the surfaces of the joint).
(b) The joint should be wide enough for the sealant to be physically applied ensuring that a continuous bond has been made. It should be wide enough for a sealant with a given flexibility to accommodate the estimated movement of the joint; this may mean extra chasing.

6 *Horizontal dpc stops at reveal, instead of extending sideways into jamb to form overlap with vertical dpc in jamb (see circled 4 in diagram 5). (Photo Building Research Establishment, Crown Copyright.)*

(c) Use of back-up strip and/or bond-breaking tape. Sealants are designed to be bonded to two surfaces only so that their movement is unrestrained except at those two surfaces. A resilient foam back-up strip ensures that the sealant does not spread into the depth of the joint but is forced against the two surfaces being sealed.
(d) Where fillets are used for pointing they should have a convex profile.

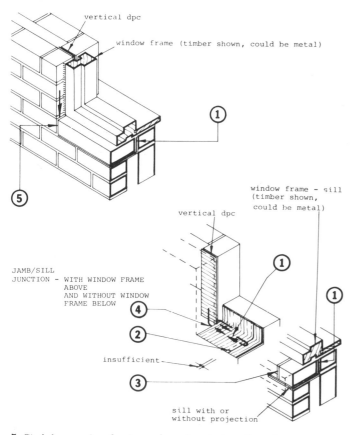

5 *Circled numerals refere to numbers given in parentheses in text.*

7 *Inadequate provision for differential movement between sill and jamb (see circled 5 in diagram 5) has resulted in joint opening up and admitting water.*

2 *At worst*:
Take out the sill and dpc – if any – and remake with correctly formed dpc and sill joints.

References for correct detailing
1 Correct sill details:
  (a) Handisyde, C. C., *Everyday details* 11, Window sills.
2 For correct dpc details:
  (a) Duell and Lawson, *Damp-proof course detailing*, Fig. 20, page 14, and Fig. 52a, page 27.
  (b) BS 5628: Part 3: 1985: *British Standard Code of Practice for use of masonry*, 'Materials and components, design and workmanship' (formerly CP 121: Part 1: 1973). Fig. 12, page 46.
  (c) BRE Report, *Thermal insulation: avoiding risks*, 2 Walls, 215, pages 16 for cold bridges and Fig 30, page 17 for other parts of the opening. Concise with excellent illustrations.
3 For selection of sealants for pointing:
  (a) BRE DAS 68, *External walls: Joints with windows and doors – detailing of sealants* (Design) has list of sealants suitable for use around windows.
  (b) Sealant Manufacturers' Conference and CIRIA: *Manual of good practice in sealant application*. Sets out the principles both of selection and application – poor illustrations.

Cause 2: not associated with rain
Surface condensation and/or mould growth is less common immediately underneath the sill, but if present it is probably due to cold bridging, as described for lintels and jambs above.

Cross reference
Study 20: Condensation under heading 'Surface condensation: cold bridges', 4.02, page 108.

Manifestation 2
Timber decay, usually externally but also internally. Both windows and doors may be affected – the former more commonly.

Cause
The decay is most commonly due to the growth of wet rot fungi on unpreserved timber (especially soft woods having a low resistance to this type of decay). The timber may have become sufficiently damp for one or more of the following reasons:

1 Because moisture was entrapped initially when the window was painted.
2 Because moisture has migrated from wet masonry or concrete, in contact with the window frame. Weaknesses at jamb/sill junctions at positions 4 and 5 on diagram **5** could be significant.
3 Because moisture has entered through open joints in the window frame itself.
4 Because of surface condensation on the inner face of the glass.

Cross reference
Study 27: Timber decay, page 163.

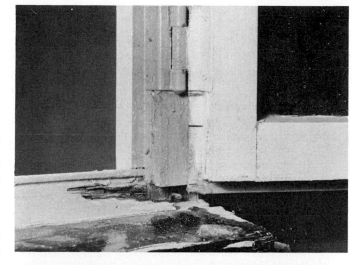

8, 9, 10 *Different locations of decay in timber windows. (Photos Building Research Establishment, Crown Copyright.)*

Remedial work: 1
1 *At best* (if decay is localised):
Cut away the affected parts and renew them with preserved timber. Allow the timber to dry and then repaint.

2 *At worst*: (if decay is extensive):

Renew the window completely.

It is important that any migration from the masonry (or concrete) to the timber frame is prevented. To do this may require remedial work to dpcs as described for 'Jambs' and 'Sills'.

References for correct detailing

1 BRE Digest 304, *Prevention of decay in external joinery*, for use where new joinery has to be specified.

2 BRE Information Paper IP 10/80, *Avoiding joinery decay by design* summarises recent research work and sets out the following guidelines to avoid decay:

  (a) Use timbers of suitable durability or employ preservative treatment.

  (b) Any dowels used should be of a durable timber or pre-treated with a water-repellent preservative.

  (c) Design to avoid water traps and horizontal surfaces.

  (d) Avoid jointed sills and bottom rails.

  (e) Seal effectively all joints between components.

  (f) Use durable glues.

  (g) Seal the edges of plywood.

  (h) Do not fit mortice locks in region of a dowelled joint.

3 BRE Digest 201, *Wood preservatives: pre-treatment application methods*. The table Classification of wood preservatives (below) is an extract.

4 BRE Digest 354 *Painting exterior wood* (replaces Digest 261). The table of general properties of paint systems below is an extract.

Table: general properties of paint systems. (Table from Digest 354) (Building Research Establishment, Crown Copyright).

Property	Conventional paints (solvent-borne)	Exterior-quality paints	
		Solvent-borne	Water-borne
Adhesion	Good over well-prepared surfaces		
Long-term extensibility	Poor	Moderate	Good
Flow	Good	Good	Moderate
Gloss levels(1)	High	High	Moderate
Colour stability	Moderate	Moderate	Good
Moisture permeability	Low	Low to moderate	Moderate to high
Maintenance interval	3–4 years	4–6 years	5–8 years
Redecoration procedure	Difficult	Moderate	Easy
Compatibility with putty	Good	Moderate	Moderate
Tolerance to adverse weather (during application)	Good	Good	Moderate
Blocking resistance(2)	Good	Good	Moderate

(1) Low-gloss level finishes available in all paint types
(2) Self-adhesion between contacting surfaces

Table: Classification of wood preservatives. (Table from BRE Digest 201) (Building Research Establishment, Crown Copyright).

Preservative types and ingredients	Treatment methods	Preservative properties
Tar oil Distillate from coal tar	1 Coal tar creosote to BS 144 — Pressure, Open tank	Resist leaching and are particularly suitable for external work. Have a characteristic odour, and can stain adjacent materials.
	2 Coal tar oil to BS 3051 — Pressure, Open tank, Immersion, Brush	Non-corrosive to metals, and treated timber presents no special fire hazard after a few months' drying. Not suitable normally for timber that is to be painted. Impart a degree of water-repellency to timber which helps in retarding dimensional movement.
Organic solvent to BS 5707: Part 1. (Solutions of one or more organic fungicides/insecticides in organic solvents, usually petroleum oil distillate)	1 Copper naphthenate 2 Zinc naphthenate 3 Pentachlorophenol and derivatives 4 Tributyltin oxide 5 gamma-HCH 6 Dieldrin — Double vacuum, Immersion, Delunging, Brush/spray	Most of the preservatives are resistant to leaching but some are subject to loss by evaporation: suitable for exterior and interior use. Not generally corrosive to metals, and non-staining. Treated timber clean in appearance and, when solvent has dried off, can usually be painted and glued satisfactorily. Treatment does not cause swelling of the timber and these preservatives can be employed on accurately machined wood and components without trouble from movement or distortion. Solvents are readily flammable, but once they have evaporated the treated timber presents no fire hazard. Water-repellent additives can be included to retard timber moisture changes in service.
Water-borne Inorganic salts dissolved in water	1 Copper/chrome/arsenic to BS 4072 — Pressure 2 Copper/chrome to BS 3452	(1) and (2) undergo chemical changes within the wood and become resistant to leaching: suitable for exterior and interior use. (3) is leachable to some extent, but can be used outside if the wood is painted. Generally non-staining and non-flammable. Copper-based preservatives may induce some metal corrosion in severe environments. Timber must be re-dried after treatment: clean (but sometimes coloured) in appearance: can be painted and glued satisfactorily.
	3 Disodium octaborate — Diffusion	

5 BRE Information Paper IP 34/79, *Exterior wood stains* concludes that in joinery stains are successful only on high quality material and if maintained every three years. Different glazing techniques are required – beads or gaskets, rather than traditional putty, or modern glazing compounds should be used.

6 BRE Digest 286, *Natural finishes for exterior timber,* discusses the selection and use of varnishes and stains.

7 BRE Technical Note 28 (revised 1979), *Maintenance and repair of window joinery,* states the causes of decay in window joinery and gives an account of the remedial measures which may be taken by a householder to prevent moisture penetration and to deal with the early stages of decay.

8 BRE DAS 13, *Wood windows: arresting decay* (Design) discusses factors leading to premature decay and their prevention and has a brief note on remedial work.

9 BRE DAS 14, *Wood windows: preventing decay* (Design) gives advice on prevention.

10 BRE DAS 15, *Wood windows: resisting rain penetration at perimeter joints* (Design).

11 BRE Digest 321, *Timber for joinery* gives information on timbers used in joinery and guidance on selection.

Remedial work: 2

If condensation has been the chief cause of decay then, apart from repairing the window frame (see above), a drainage channel should be incorporated that will allow the condensation to drain to the outside. References to details are given below. In situations where there is heavy moisture production, more ventilation and/or heat input may be necessary in addition.

11 *Decay in door bottom rail; timber exposed to damp should always be given preservative treatment (softwood especially).*

Note

For provision of a drainage channel in a timber window sill to reduce the risk of wood decay (From Handisyde, C. C. Everyday details, 11 Window sills, page 52), see page 107.

References for correct detailing

1 DOE, *Condensation in dwellings part 2.*

2 Handisyde, C. C., *Everyday details,* 11 gives more comprehensive details on page 52.

Study 17
Claddings

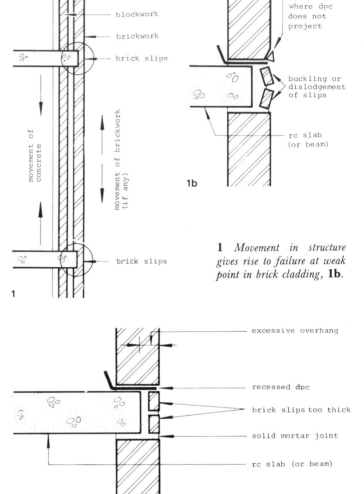

1 *Movement in structure gives rise to failure at weak point in brick cladding,* **1b**.

E 2,3,4

Manifestation 1
Buckling and/or dislodgement of brick slips; with or without spalling of the bricks immediately above the slips.

Condition
The schematic diagram **1**, top right, shows the constructional detail in which the above manifestation of failure has arisen. It is a generalized detail, and the principles illustrated must be interpreted in relation to the specific construction found on site in any particular case.

Cause
Essentially moisture movement. The cladding is literally squeezed by the vertical shrinkage and creep of the concrete frame but this may be accentuated by the expansion of the bricks. Internal blockwork is normally not damaged.

The tight joint, particularly below the rc slab (or rc beam), fails to provide pressure relief, by not allowing any movement to take place.

Cross references
Study 24: Movements under headings 'Basic mechanisms', 4.02, page 136, 'Moisture movements', 8.01, page 140, and especially under 'Movement joints', 8.05, page 144.

Remedial work
1 *At best* (if the cladding above the slab or beam is structurally safe and most of the movement has already taken place):

Remove the damaged brick slips. Cut out the mortar joint below the rc slab or rc beam and fill the joint with a sealant backed with a foamed plastic (both the sealant and the plastic foam must be capable of accepting the compressive forces induced by any further movement). If necessary, extend the dpc (to ensure that it is not recessed behind the wall face), using a metal tray underneath the existing dpc and extending beyond the outer face of the cladding (for the insertion of the metal tray see Study 14, Manifestation 4, page 81). Refix the brick slips using waterproofing/adhesive additive such as Ronafix.
2 *At worst* (if the cladding is structurally unsafe):

Rebuild part or all of the cladding incorporating a pressure relief joint as described above and detailed in the references given below.

References for correct detailing
1 General:
   (a) **BRE** Digest 223, *Wall cladding: designing to minimize defects due to inaccuracies and movements*, contains a detailed discussion of the types of movement affecting cladding and how to design fittings and joints to accommodate it. There are no special constructional details.
   (b) **BRE** Digest 217, *Wall cladding defects and their diagnosis*. describes a clear methodology for observing and

2 *Typical structural condition in which failure may occur.*

3 *Typical example of dislodged brick slips at concrete slab. (Photo by courtesy of Brick Development Association.)*

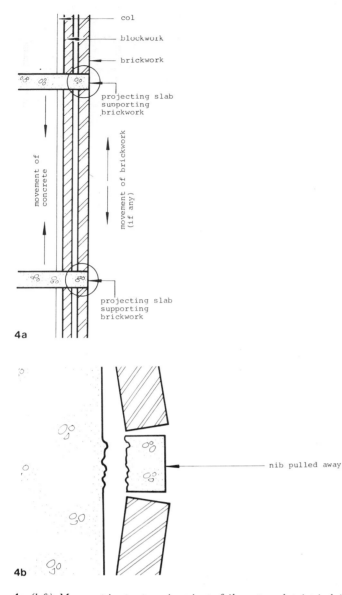

4a (left) Movement in structure gives rise to failure at weak point in brick cladding, **4b** (right).

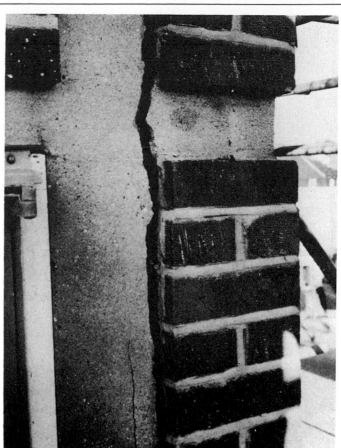

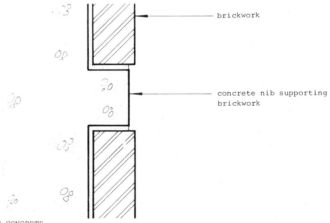

5 Dislodgement of brick cladding and supporting concrete ledge (Photo Building Research Establishment, Crown Copyright).

6 Typical structural condition in which failure may occur.

diagnosing non-loadbearing cladding failures, including brick slips.

(c) BRE DAS 2, *Reinforced-concrete frame flats: repair of disrupted brick cladding*, explains mechanism of failure and remedial work required.

(d) GLC, *Development and Materials Bulletin* **101**, January 1977, item 2, 'Fixing of brick slips – feedback on failure'. This is a report of a survey of brick slips fixed to vertical faces of reinforced concrete using styrene butadiene rubber emulsion gauged cement and sand mortar. Success or failure depended largely on good workmanship and compliance with the specification.

2 For pressure relief joints:

(a) Brick Development Association *Technical Note 9*, April 1975, Figs 5–8 on page 3.

(b) DOE, *Construction 14*, page 17, Fig. 7 (reproduced in Study 24 on page 194).

Manifestation 2

Buckling and/or dislodgement of brick cladding with or without dislodgement of the supporting concrete.

Condition
The schematic diagram **4** shows the constructional detail in which the above manifestation of failure has arisen. It is a generalized detail, and the principles illustrated must be interpreted in relation to the specific construction found on site in any particular case. The cause of the failure is as described for Manifestation 1 above.

Cross references
Study 24: Movement under headings 'Basic mechanisms', 4.02, page 137, 'Moisture movements', 8.01, page 141, and especially under 'Movement joints', 8.05, page 144.

7–12 *Examples of dislodgement of tiling, mosaic, rendering, and brick slips. (Photos GLC.)*

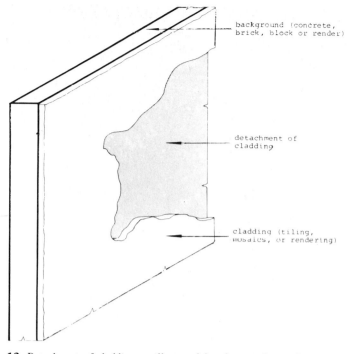

background (concrete, brick, block or render)

detachment of cladding

cladding (tiling, mosaics, or rendering)

13 *Detachment of cladding as illustrated by photographs on the previous page).*

Remedial work

Generally rebuild part or all of the cladding, incorporating a pressure relief joint as described below. If, however, the concrete nib has been dislodged, a specialist firm will then have to advise on methods of providing support for the cladding and providing adequate cover to the reinforcement. Epoxy resins have been used successfully in repairing broken concrete.

1 *At best* (if the cladding above the slab or beam is structurally safe and most of the movement has already taken place):

Renew the damaged brickwork. Cut out the mortar joint below the rc slab or rc beam and fill the joint with a sealant backed with a foamed plastic (both the sealant and the plastic foam must be capable of accepting the compressive forces induced by any further movement). If necessary, extend the dpc (to ensure that it is not recessed behind the wall face), using a metal tray underneath the existing dpc and extending beyond the outer face of the cladding (for the insertion of the metal tray see Study 14, Manifestation 4, page 81).

2 *At worst,* (if the cladding is structurally unsafe):

Rebuild part or all of the cladding incorporating a pressure relief joint as described above and detailed in the references given below.

References for correct detailing

1 Brick Development Association *Technical Note 9*, April 1975, Figs 5–8, page 3.

2 DOE, *Construction* **14**, page 17, Fig. 7 (reproduced in Study 24, page 144).

3 BS 5638: Part 3: 1985: *British Standard Code of Practice for the use of masonry*, 27.3 Brick and block slips, page 71 and Fig. 13, page 72

Manifestation 3

Cracking and/or detachment of tiling, mosaic or rendered finishes.

Condition

The schematic diagram **13** shows the constructional detail in which the manifestation of failure, **3**, has arisen. It is a generalized detail, and the principles illustrated must be interpreted in relation to the specific construction found on site in any particular case.

Cause

Can be differential movement, chiefly from moisture movement, between the finish and the background, especially in the absence of movement joints (or of adequately spaced movement joints, in the case of tiling and mosaics).

With tiling and mosaics, the cause of failure may be entirely due to the use of the wrong adhesive or the incorrect use of the correct one.

With a rendered finish, sulphate attack may be responsible for the failure.

Cross references

Study 24: Movement under heading 'Moisture movements of materials', 8.01, page 140.

Study 25: Loss of adhesion under heading 'External render', 7, page 151 and 'External wall tiling and mosaic', 9 page 153.

Remedial work

1 *At best* (if cracking or detachment is not too extensive and only a small amount of differential movement can still be expected):

Local repair of the affected area(s).

2 *At worst* (if cracking or detachment is extensive; and/or adhesion is suspect):

(a) Complete removal and reinstatement of the finish. Movement joints will be necessary if there is likely to be further shrinkage of the background. This shrinkage is a form of moisture movement and can be estimated by measuring the background's moisture content and comparing this with known values of the material in its 'dry state'. Guidance is given in Study 24 on techniques and data for estimating moisture movement. See below for other references.

(b) Overcladding with or without the removal of the defective finish, depending on the stability of the latter and/or the background to which it was bonded.

References for correct detailing

1 For rendered finishes:

(a) BS 5262: 1976: *Code of Practice for external rendered finishes.*

See in particular paragraph 51 on page 41, 'Repair of damaged areas', which recommends that, apart from using a compatible render in accordance with the code, the use of bonding agents is advisable.

(b) C&C, *Appearance matters 2 External rendering*, 1982, is a good backup/explanation of BS 5262.

2 For tiling and mosaics:

BS 5385: Part 2: 1978: *Code of Practice for wall tiling: External ceramic wall tiling and mosaics.* Tables 2 and 3 on pages 19 and 20 give details of the correct adhesive to be used on a particular background

Note: Unlike CP 212: Part 2: 1966: *External ceramic wall tiling and mosaics* that it supersedes, this code gives guidance on a wide range of adhesives, movements and the effects of exposure.

3 For overcladding:
  (a) BS 8200: 1985: *British Standard Code of Practice for design of non-loadbearing external vertical enclosures for buildings*, deals with functional requirements and principles. No details.
  (b) BRE Report, *Overcladding external walls of large panel systems* by H. W. Harrison, J. H. Hunt and J. Thomson, addresses all the main issues and has case studies. The principles are equally applicable to other forms of construction.
  (c) 'Light cladding systems' by Barry Josey, *The Architect* (now returned as the *RIBA Journal*), September 1987, pages 61–81, is an excellent chatty review of what is involved and what is available.
  (d) Brookes, Alan, *Cladding of buildings* is a good review of a wide range of claddings with details.
  (e) Anderson, J. M. and Gill, J. R., *Rainscreen cladding, a guide to design principles and practice*, CIRIA/ Butterworths, 1988. Good for principles; more experience needed in practice.

Manifestation 4

Bowing and/or cracking of tiled or mosaic finishes bonded to timber, timber-based or other organic boards with the composite fixed to brickwork or concrete.

Condition

The schematic diagram **14** shows the constructional detail in which the manifestation of failure, **4**, has arisen. It is a generalized detail, and the principles illustrated must be interpreted in relation to the specific construction found on site in any particular case.

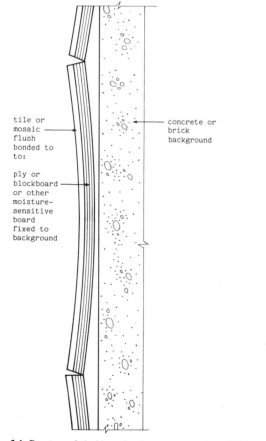

14 *Bowing of cladding fixed as composite board (Manbifestation 4).*

tile or mosaic flush bonded to to:

ply or blockboard or other moisture-sensitive board fixed to background

concrete or brick background

15 *Detachment of finish from timber substrate: see below for possible causes. (Photo Building Research Establishment, Crown Copyright.)*

Cause

Moisture movement of the board that is unrestrained at the back (i.e. because the composite is unbalanced).

Cross references

Study 24: Movment under '8 Moisture movement', page 141.

Remedial work

Removal ofthe composite and replacement with balanced composite board or other type of cladding.

References for correct detailing

None. The principle of balance embodied in plywood/ blockboard practice of having an unequal number of plies needs to be followed.

Manifestation 5

Blistering of paintwork or
Deterioration of clear varnish treatment or
Decay of timber framing or
Corrosion of zinc coated steel fastenings.

Cause

1 Blistering of paintwork because of entrapment of moisture; inappropriate primer and/or inappropriate preparation of the surface of the timber, either or both being incompatible with the paint system.
2 Deterioration of the varnish because of the short life span of clear varnishes (about two years) on exposure to external atmospheric conditions.
3 Decaying of timber because unpreserved timber was used in a location subjected to dampness.
4 Corroding of fastenings because of the action of copper-chrome-arsenate salts-based timber preservative under damp conditions.

Cross references

Study 27: Timber decay, page 162.
Study 26: Corrosion under heading 'Corrosion in fixings'. 4. page 159.

Remedial work

For paintwork and varnish: strip, prepare the timber and apply the correct paint system.

For decay: cut out the affected parts, treat adjacent timber with preservative and replace with preserved timber.

For corrosion: replace all the fastenings that can be protected or are made of a material which will inherently resist corrosion.

References for correct detailing
1 BRE Digest 354, *Painting wood* (replaces Digest 261).
2 BRE Digest 304, *Preventing decay in external joinery.*
3 GLC, *Development and Materials Bulletin* **91**, January 1976, item 5, 'Corrosion of zinc coated steel fasteners in conjunction with preservative treated timber'. This is a report of a thorough study of the effect of a wide range of timber preservatives on galvanized steel. Those based on copper-chrome-arsenate salts are the most aggressive but others should be used with caution.

Manifestation 6

Dampness appearing on the inside of cladding; and/or on adjacent floor or wall surfaces.

Condition

The schematic diagram **16** shows the constructional detail in which the above manifestation of failure has arisen. It is a generalized detail, and the principles illustrated must be interpreted in relation to the specific construction found on site in any particular case.

Cause

Dampness from leakage of rainwater through fixing holes (1 on the diagram), or around clips, or through the joint (2 on the diagram). The reason could be that the joint was not properly or fully filled; or that the adhesion of the sealant was poor (loss of adhesion may have been due to thermal movement); or that there was no side packing.

Cross reference

Study 25: Loss of adhesion under heading 'Sealants at joints, walls and around windows', 8, page 152.

Remedial work

A specialist firm is required to advise on and execute the sealant application.

Reference for correct detailing
1 For claddings generally:
  (a) Rostron, R. M., *Light cladding of buildings*, London, Architectural Press, 1964
  (b) Brookes, A. J., *Cladding of buildings*, London and New York, Construction Press, 1983.

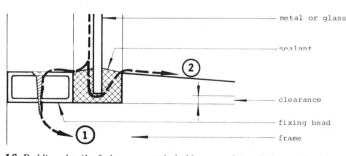

16 *Bedding detail of glass or metal cladding panel in which manifestations of failure 6 to 8 might occur.*

  (c) Brookes, A. J. and Grech, C. *The building envelope, applications of new technology cladding*, Butterworth Architecture, 1990. An excellent survey with 33 detailed case studies.
  (d) BS 8200: 1985: *British Standard Code of Practice for the design of non-loadbearing external vertical enclosure of buildings.* Good for principles.
  (e) March, P., *Fixings, fastenings and adhesives*, London and New York, Constructional Press, 1984.

Manifestation 7
Corrosion of mild steel fixings

Condition

The schematic diagram shows the constructional detail in which the above manifestation of failure has arisen. It is a generalized detail, and the principles illustrated must be interpreted in relation to the specific construction found on site in any particular case.

Cause

Corrosion of mild steel fixings due to inadequate protection of the steel for the degree of exposure to dampness – the latter would include undrained condensation behind the panel.

Cross reference

Study 26: Corrosion under heading 'Corrosion in fixings', 4, page 159.

Remedial work

1 *At best* (if the corrosion has not affected the structural safety of the fixing):
  Remove the rust and treat the metal as described in Study 26 (reference above).
2 *At worst* (if the corrosion has affected the structural safety of the fixing):
  Replace the steel fixings with non-ferrous metal fixings making sure that there is no possibility of electrolytic action as described in Study 26, 2.02, page 156.

References for correct detailing
For claddings generally:
  (a) Rostron, R. M., *Light cladding of buildings*, London, Architectural Press, 1964.
  (b) Brookes, A. J., *Cladding of buildings*, London and New York, Construction Press, 1983.
  (c) Brookes, A. J. and Grech, C. *The building envelope, applications of new technology cladding*, Butterworth Architecture, 1990. An excellent survey with 33 detailed case studies.
  (d) BS 8200: 1985: *British Standard Code of Practice for the design of non-loadbearing external vertical enclosure of buildings.*
  (e) Anderson, J. M. and Gill, J. R., *Rainscreen cladding, a guide to design principles and practice*, CIRIA/ Butterworths, 1988. Good for principles; more experience needed in practice.
  (f) March, P., *Fixings, fastenings and adhesives*, London and New York, Constructional Press, 1984.

Manifestation 8
Distortion of metal infill panels or cracking of glass.

Condition 1

The schematic diagram shows the constructional detail in which the above manifestation of failure has arisen. It is a generalized detail, and the principles illustrated must be interpreted in relation to the specific construction found on site in any particular case.

Cause
Distortion of the metal infill panel or cracking of glass due to inadequate clearance around the panel or pane for the colour of the material (dark materials absorb more heat and expand more than very light ones) and the degree of exposure (to sunlight especially). Note that the cracking of the glass may also be due to the corrosion of mild steel frames and beads – the corrosive product occupies or fills spaces intended for movements.

Cross reference
Study 24: Movement under 7, page 138.

Remedial work
Where there is distortion and cracking, the panel will have to be removed and replaced with a new panel with sufficient clearance to suit the material and the degree of exposure to sunlight. This can be calculated knowing the size of the

panel, its linear coefficient of thermal expansion and the likely temperature range, the latter depending on orientation. As a guide, a minimum clearance of 5 mm at each edge for panels with a dimension exceeding 750 mm is normally recommended.

If there is corrosion, the rust will have to be removed and the metal suitably treated before replacing the panel.

Condition 2
The schematic diagram **17**, shows the constructional detail in which the manifestation of failure, **4**, has arisen. It is a generalized detail, and the principles illustrated must be interpreted in relation to the specific construction found on site in any particular case.

Cause
Distortion of the metal facing due to differential thermal movement between the metal facing and the thermal insulation bonded to it. The steep temperature gradient across the insulant aggravates the problem, as does the absence of a balancing metal facing bonded to the insulant (i.e. there is nothing to restrain the movement of the insulant). The problem is increased if the metal facing is of a dark colour and is relatively weak (e.g. aluminium compared to the stronger steel).

Remedial work
1 *At best:* If the distortion of the metal is simply a matter of unacceptable appearance, there is no need for remedial work provided that condensation is not occurring between the insulant and the metal facing.
2 *At worst:* Remove the cladding and replace with cladding where the insulant sandwiched between two sheets of metal or where the insulant is completely detached from the metal facing.

References for correct detailing
1 For cladding generally:
   (a) Rostron, R. M., *Light cladding of buildings*, London, Architectural Press, 1964.
   (b) Brooks, A. J., *Cladding of buildings,* London and New York, Construction Press, 1983.
   (c) Brookes, A. J. and Grech, C. *The building envelope, applications of new technology cladding*, Butterworth Architecture, 1990. An excellent survey with 33 detailed case studies.
   (d) BS 8200: 1985: *British Standard Code of Practice for design of non-loadbearing external vertical enclosure of buildings.*
   (e) Anderson, J. M. and Gill, J. R., *Rainscreen cladding, a guide to design principles and pracctice*, CIRIA/ Butterworths, 1988. Good for principles; more experience needed in practice.
2 For glass in particular:
   BS 6262: 1982: *British Standard Code of Practice for glazing and fixing of glass for building.*

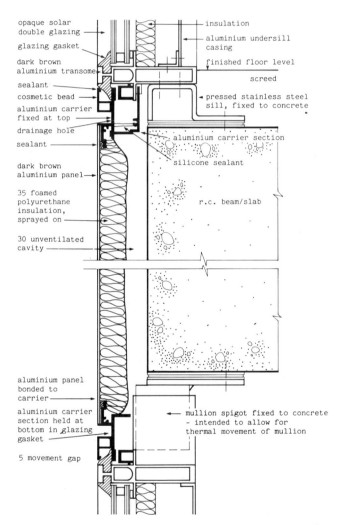

opaque solar double glazing
glazing gasket
dark brown aluminium transome
sealant
cosmetic bead
aluminium carrier fixed at top
drainage hole
sealant
dark brown aluminium panel
35 foamed polyurethane insulation, sprayed on
30 unventilated cavity
aluminium panel bonded to carrier
aluminium carrier section held at bottom in glazing gasket
5 movement gap

insulation
aluminium undersill casing
finished floor level
screed
pressed stainless steel sill, fixed to concrete
aluminium carrier section
silicone sealant
r.c. beam/slab
mullion spigot fixed to concrete – intended to allow for thermal movement of mullion

17 *Diagrammatic section of cladding with bonded polyurethane insulation on the back (Manifestation 7, Condition 2).*

Study 18
Wall/ground floor junctions

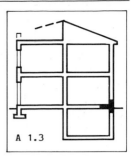

A 1.3

Manifestation

Persistent rising damp internally at or near the floor level, with or without mould growth. Where dampness is intermittent and connected with cold spells condensation may be responsible.

Condition

The schematic diagram **1** shows the constructional details in which the above manifestation of failure may have arisen. The details are generalised and the principles illustrated must be interpreted in relation to the specific construction on site.

Causes

Rising ground moisture (but see note about existing buildings after item 6 below) due to one or more of the following:

Bridging due to the lack of continuity between the dpc in the wall and the dpm in the floor, causing rising dampness in the following numbered positions on the digram:

1 wall
2 screed — leading often to the loss of adhesion of the floor finish.
3 plaster — this could be the sole means by which bridging has taken place.
4 breakdown of the dpm due to puncturing of the thin dpm material by roughness of the substrate; or by discontinuity of a 'continuous' (cold applied) membrane; or by inadequate laps in sheet material. In existing buildings which have been damp-proofed (see diagram) moisture entry may be due to bridging by either:
5 external rendering; or
6 splashing of rainwater where the dpc is too low.

Note:

In existing buildings especially, dampness may also be accompanied by mould growth. Condensation and/or hydroscopic salts may be contributing to the dampness. Condensation is a risk during cold spells when the heating is too low and ventilation inadequate to prevent a build-up of moisture in the building. Mould growth is a risk when the relative humidity of the internal air is above 70 per cent for long periods (i.e. more than 12 hours per day). These aspects should not be overlooked — see the cross references below.

Cross references

Study 20: Condensation, page 106.
Study 23: Rising damp, page 131.

Remedial work

1 *At best* (if rendering or plaster is the sole cause of bridging):

Remove the render or plaster from at least 25 mm above the dpc down to ground level or screed level. Internally, this may require a higher skirting than the existing one. The new skirting (or the existing skirting, if re-used) should be treated with preservative and the back of the skirting should be primed and painted before fixing. If splashing of rainwater is the sole cause because the dpc is too low, lower the ground level.

2 *At worst:*

Partial opening up will be required to see whether there is in fact discontinuity between the dpm in the floor and the dpc in the wall. There is no standard method for making a link remedially but the following methods could be tried.

Inject silicone at skirting level. As with walls this should be done at a dry period so that the silicone solutions remain at the critical point. Alternatively, cut a vertical slot along the wall/floor junction so that both the dpc and the dpm are exposed. Make good with waterproof cement:sand screed or sealant that is compatible with both the dpc and the dpm materials.

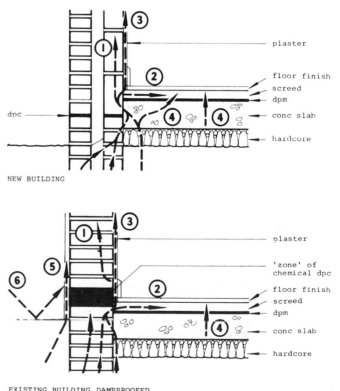

NEW BUILDING

EXISTING BUILDING DAMPPROOFED
(CONC SLAB REPLACING EXISTING SUSPENDED TIMBER FLOOR)

1 *Penetration paths in new work (top and middle above), and in existing buildings (below) subsequently damp-proofed (above). Circled numbers refer to notes in text.*

References for correct detailing

1 For new buildings:

   (a) Handisyde, C. C., *Everyday details* 5, 'Masonry walls: dpc at base of external walls'.

   (b) CP 102: 1973: *Protection of buildings against water from the ground*. The relevant drawings are unclear but clause 11.4 states:

   'It is essential that the dpm in the floor should be continuous with the dpc in the surrounding walls'.

   (c) BRE Digest 54, *Damp-proofing solid floors* tabulates the range of possible materials for dpms and the resistance to dampness of different floor finishes.

   (d) BRE DAS 35, *Substructure: DPCs and DPMs – specification* gives essential information with good illustration of polythene dpm/dpc junction.

   (e) BRE DAS 36, *Substructure: DPCs and DPMs – installation* gives essential information with helpful illustrations of dealing with laps.

2 For existing buildings:

   (a) BRE Digest 245, *Rising damp in walls: diagnosis and treatment*, January 1981 (supersedes Digest 27, *Rising damp*) and incorporates fuller information on chemical dpcs and the more accurate measurement of moisture content using the BRE drilling method.

   (b) Duell and Lawson, *Damp-proof course detailing*. Fig. 13, page 10 for correct detailing and chapter 9, pages 35–41, for remedial work.

   (c) Agrément certificates for chemical dpcs.

   (d) Code of practice for the insertion of chemical dpcs, the British Chemical Dampcourse Assocation.

   (e) BS 6576: 1985: *British Standard Code of Practice for installation of chemical damp-proof courses*.

   (f) BRE DAS 85, *Brick walls: injected dpc's*.

   (g) BRE DAS 86, *Brick walls: replastering following dpc injection*.

   (h) BRE GBG 3, *Damp proofing basements*.

Study 19
Floors and ceilings

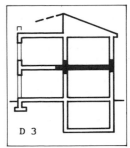

Manifestation 1
Rippling of thin flooring.

Condition
The schematic diagram **1** shows the constructional detail in which the above manifestation of failure has arisen. It is a generalised detail, and the principles ilustrated must be interpreted in relation to the specific construction found on site in any particular case.

Cause
Moisture movement of the screed (shrinkage and expansion). The edges of the screed at construction joints or cracks curl on drying, lifting the floor finish. Subsequently the screed expands on absorbing moisture and flattens. However, the previously expanded floor finish does not flatten 'in sympathy' but instead it ripples.

Cross references
Study 24: Movement under headings 'Basic mechanisms, tensile and compressive stresses', 4.02, page 136, and 'Drying out', 8.03, page 142.

Remedial work
Cut out the rippled flooring and fill the gap in the screed to its full depth with a material capable of resisting screed movements. Finally reinstate the flooring.

Suitable filling materials might be a 3:1, sand:cement mix (by weight) for gaps wider than about 30 mm well compacted; or a 6:1 thermosetting resin:sand mix (by weight) for narrower gaps.

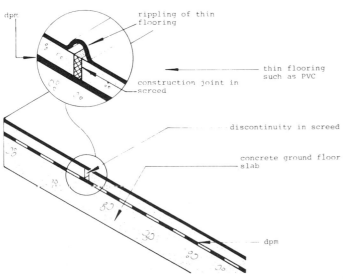

GROUND FLOOR

1 *Constructional detail in which failure shown at left may have arisen.*

2, 3 *Examples of thin floorings. (Building Research Establishment, Crown Copyright.)*

Reference for correct detailing
BRe Current Paper CP 94/74, *The rippling of thin flooring over discontinuities in screeds* by W. J. Warlow and P. W. Pye (page 7 for preventative measures).

Manifestation 2
1 Lifting of thermoplastic and PVC (vinyl) tiles;
2 Blistering and/or loss of adhesion of thin sheet flooring materials.

Condition
The schematic diagram **4** shows the constructional details in which the above manifestation of failure has arisen. These are generalised details, and the principles illustrated must be interpreted in relation to the specific construction found on site in any particular case.

Cause
Rising dampness due to ineffective damp-proofing (such as splits in the membrane or poor lap joints) or to the absence of any damp-proofing at all. Dampness may attack the adhesive (alkali from the cement may be significant) or cause dimensional changes in the flooring material.
    Note: If the failures are localised near external walls dampness may be due to lack of continuity between the dpc in the wall and the dpm in the floor — see Study 18 earlier.

Cross references
Study 23: Rising damp, Fig 3 on page 132 is extracted from CP 102 (see below) and gives details of the effect of dampness on different floor finishes.
    Study 25: Loss of adhesion under heading 'Impervious sheet floor finishes', 13, page 155.

Remedial work
1 At best (if an incorrect adhesive was used and there is no dampness):

    Take up the existing flooring and relay with the correct adhesive following the advice of the adhesive manufacturer.

2 *At worst* (if there is dampness and this is the major cause of the failure):
    If it is not possible to incorporate a dpm, take up the flooring and relay with flooring that is not moisture sensitive. Examples of the latter are pitch mastic or mastic asphalt flooring; concrete terrazzo or clay tile flooring; cement/rubber latex or cement/bitumen flooring; or wood blocks dipped and laid in *hot* pitch or bitumen.

References for correct detailing
1 CP 102: 1973: *Protection of buildings against water from the ground*, table 2, page 23.
2 BRE Digest 54, *Damp-proofing solid floors*, Table 1, page 2.

Manifestation 3
Dampness on the ceiling (or the floor) at or near the junction with the external wall.

Condition
The schematic diagram **5** shows the constructional details in which the above manifestation of failure has arisen. These are generalised details, and the principles illustrated must be interpreted in relation to the specific construction found on site in any particular case.

Cause
1 If not associated with rain:
    Cold bridging caused by the concrete has resulted in surface condensation. The cold bridge effect is likely to be worse if the slab projects.

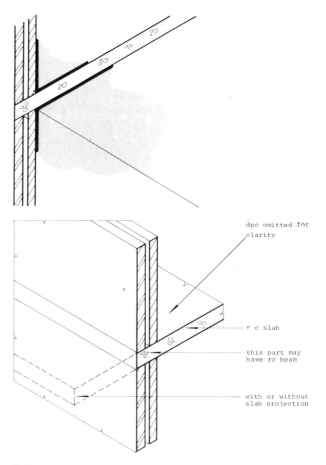

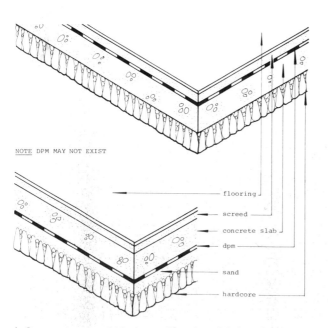

NOTE DPM MAY NOT EXIST

flooring
screed
concrete slab
dpm
sand
hardcore

4 *Constructions in which the manifestation of failure described above may have arisen.*

dpc omitted for clarity

r c slab

this part may have rc beam

with or without slab projection

5 *Constructional situations in which manifestation of failure described above may have arisen (dark tone represents damp).*

6 *Example of cracking of partition wall on slab. (Photo Building Research Establishment, Crown Copyright.)*

2 If associated with rain:
There has been water penetration due to faulty dpcs in the wall. This aspect is dealt with fully under Study 14: RC frame/cavity wall junctions, Manifestations 3, 4 and 5, pages 80–82.

Cross reference
Study 20: Condensation under heading 'Surface condenstion: cold bridges', 4.02, page 108.

Remedial work
1 Improve the thermal insulation locally by applying an adequate thickness of extruded polystyrene. This has the advantage over other thermal insulants in having the properties of a vapour check. If other thermal insulants are proposed which are vapour permeable, such as expanded polystyrene, a vapour control layer on their warm side will be required. Composite panels of extruded polystyrene and plasterboard (Styroliner) and expanded polystyrene, polythene and plasterboard (Gyproc) are now available. The thickness of the insulant will probably not be less than 15 mm. The minimum amount required can be calculated following the methods described in the references listed at the end of Study 20: Condensation, 5, page 112. The thermal insulation will have to be applied for a distance of 450 mm along the ceiling or floor and for a distance of at least 300 mm on the wall. It is most important that the insulant is fully bonded to the wall, floor, or ceiling. The joints between boards must be sealed with a sealant compatible with either the polystyrene or other plastics which may be used in composite boards.
2 Apart from or in addition to the improvement of thermal insulation locally, consideration should also be given to:

  (a) The improvement of ventilation; and/or
  (b) A change of the pattern of the running of the heating system; and/or
  (c) The provision of additional heating.

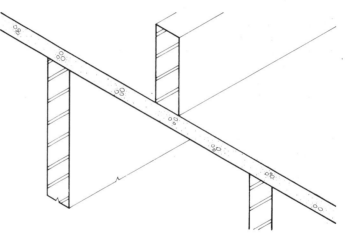

7 *Construction detail in which failures of above type may have arisen.*

References for correct detailing
1 DOE/HMSO, *Condensation in dwellings Part 2 Remedial measures*. See particularly the remedial treatment for different rooms on pages 26 and 27.
2 BS 5250: 1975: *Code of basic data for the design of buildings: the control of condensation in dwellings*. See clause 31, 'Remedial action', page 17.

Manifestation 4
Condition
The schematic diagram **7** shows the constructional detail in which the above manifestation of failure has arisen. It is a generalised detail, and the principles illustrated must be interpreted in relation to the specific construction found on site in any particular case.

Cause
Deflection due to moisture movements resulting from the use of shrinkable aggregates. This problem is confined to Scotland.

Cross reference
Study 24: Movement under heading 'Drying out, shrinkable aggregates', 8.03, page 143.

Remedial work
1 *At best:*
   Fill gaps and then redecorate.
2 *At worst:*
   Rebuild the wall providing for movement by using a soft joint (i.e. a weak mortar, 1:6, cement:sand) or a flexible joint made of a foamed plastic filling with a sealant, (e.g. a cross-linked polyethylene foam compressed 25 per cent with 2 parts polysulphide sealant).

Reference for correct detailing
BRE Digest 35, *Shrinkage of natural aggregatres in concrete* (new edition, 1971).

Manifestation 5
Separation (partial or complete) of the plaster from its substrate.

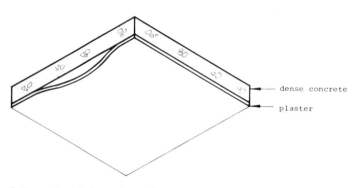

dense concrete
plaster

8 *Separation of plaster from substrate.*

Condition
The schematic diagram **8** shows the constructional detail in which the above manifestation of failure has arisen. it is a generalised detail, and the principles illustrated must be interpreted in relation to the specific construction found on site.

Causes
1 Inadequate control of the suction of the concrete background that has resulted in a weak bond; and
2 Differential movement, usually from thermal changes, causing shearing forces to be set up at the interface of the two materials sufficient to break the weak bond.

Cross reference
Study 25: Loss of adhesion under heading 'Plastering on dense concrete/soffits', 6, page 151.

Remedial work
Strip off the existing plaster and apply new plaster as for new work, making allowance for the background's characteristics. A bonding treatment such as PVAC emulsion may be used, following the manufacturer's advice, otherwise backing to make a good key is essential. Moderate wetting of the concrete improves the adhesion. The thinner the plaster coat, the smaller the shear forces set up by differential movement betwee the plaster and the background.

References for correct detailing
1 BS 5492: *Code of Practice for internal plastering*. See particularly clause 24, Page 9, 'Preparation of background'.
2 BRAS Technical Information Leaflet 24, 'Plastering on dense concrete'.
3 BRE Digest 213, *Choosing specifications for plastering*, sets out the recommendations of BS 5492 above in terms of a decision-making sequence.

Part 3: Technical studies: Causes and mechanisms of failure

Study 20
Condensation

1 Occurrence

Most commonly in dwellings (more frequently in local authority dwellings than in the private sector) but also in sheeted industrial roofs, on internal surfaces or within the thickness of a construction (flat roofs notably but not exclusively). Intense condensation has also been experienced in swimming pool roofs but these are not covered in this study, which deals only with the most commonly occurring instances.

2 Definition and mechanism

2.01 What is condensation?

Condensation is a change of water vapour in the atmosphere into liquid water that begins to take place when the content of water vapour in the air becomes equal to the maximum content of water vapour that the air can hold *at that temperature*. The liquid water precipitated is the excess water vapour that air cannot hold when it is cooled below the dewpoint of the air.

2.02 How does condensation occur?

More simply stated, when warm moist air meets a cold surface it is cooled and gives up some of its moisture as condensation, because cool air cannot hold as much water vapour as warm air – the higher the temperature of the air, the more water vapour it can hold, as shown by the psychometric chart, **1**. Air is said to be 'saturated' when it holds the maximum amount of moisture that it can contain at that temperature. When air which is not saturated is cooled, it reaches a temperature at which it does become saturated and this temperature is known as 'dewpoint'. Any further cooling of the air below dewpoint will result in the

1 *Facsimile reproduction from BS 5250: 1989 of a psychrometric chart and its use. If the dry bulb air temperature is 0°C and the moisture content is 3.5 g of moisture per kg of dry air (= 0.55 kPa vapour pressure), the relative humidity is seen to be 90%, at A (this could be the condition of outside air in winter). If that air enters a building and is heated to 20°C, relative humidity falls to 23%, B. If, however, the moisture content is raised to 10.3 g/kg (= 1.65 kPa vapour pressure) by breathing, cooking or washing activities within the building, the rh would be 70%. The rh rises to 100% (i.e. dewpoint/saturated vapour pressure) and condensation begins to occur, if the air temperature falls to 14.5°C (by meeting a cold wall surface for example).*

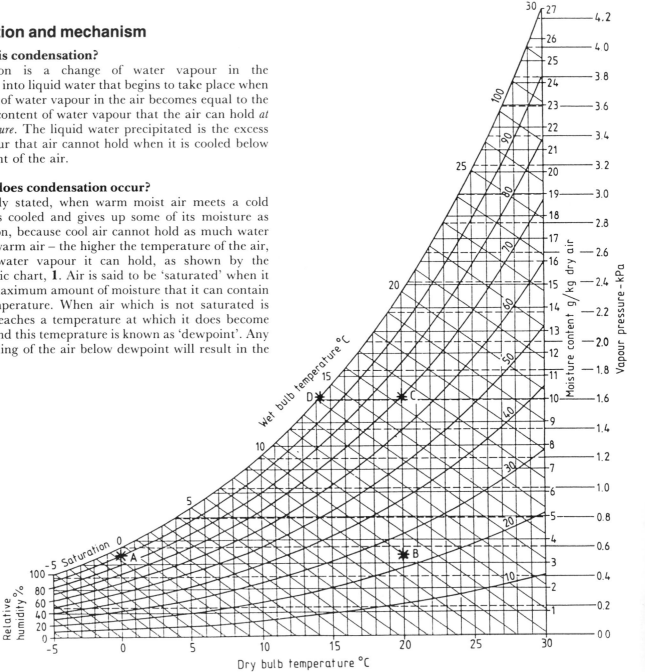

air giving up its excess moisture as condensation. 'Relative humidity' (rh) expresses as a percentage the ratio between the amount of moisture in air at a given temperature, and the maximum amount of moisture that air can hold at that temperature. Saturated air has a relative humidity of 100 per cent. Air at 20°C, 30 per cent rh has the *same* amount of moisture as air at 5°C, 80 per cent rh, but the fomer can 'absorb' considerably more moisture than the latter; air at 20°C, 70 per cent rh contains about two and a half times as much moisture as air at 5°C, 70 per cent rh.

2.03 Where does condensation occur?
Condensation may occur not only on visible surfaces (known as 'surface condensation') but also within the thickness of a construction (known as 'interstitial condensation'), diagram **2**. Whether or not condensation will occur depends on the amount of moisture in the air in contact with a surface; the temperature of this air; and the temperature of the surface. (For condensation to occur the temperature, whether on a visible surface or within the thickness of a construction, must be *below* dewpoint*.) The amount of moisture in air can be controlled by ventilation; the temperature of the air can be increased by heating (and reduced by ventilation) and the temperature of surfaces is dependent basically on thermal insualtion but is in practice also dependent on the thermal response of the building fabric – i.e. speed of heating up and cooling down in response to ambient temperature changes.

2.04 Mould growth
In contrast, mould growth (the formation of growths mainly by fungi and bacteria) is essentially moisture dependent, as the other two ingredients necessary for growth (a source of infection and a source of nourishment) cannot normally be avoided in buildings. Mould growth usually signifies the presence of moisture due to condensation.

There are many varieties of mould spores; some can germinate at relative humidities as low as 80–85 per cent.

The moulds will spread if the relative humidity is over 70 per cent for long periods (i.e. usually longer than 12 hours' duration). Apart from providing particular nourishment for the moulds, the extent to which surfaces can absorb and retain moisture can be important.

3 State of knowledge

3.01 Simple rules not reliable
The principles of condensation, known for a long time, are relatively simple. However, changes in living habits and innovations in construction have strained qualitative and 'simple rule' procedures (although if these were rigorously applied, as they obviously have not been, without question many failures could have been avoided). From a quantitative point of view, currently available methods of prediction still rely on steady-state conditions* and so the significant effects of the dynamic thermal response of the fabric cannot be taken into account reliably. They must nevertheless be allowed for. More information is required on ventilation rates and the vapour diffusing properties of a wide range of materials, apart from microclimatic data. Nevertheless, suitably interpreted (and the interpretation is very important), current methods of estimation of condensation risk do provide valuable, if not fully reliable guidance.

3.02 Use latest data
It is essential that the latest data only are used. The concept of moisture gain analysis, was introduced by developing further work by Seiffert in West Germany (for example, his *Damp diffusion and buildings*, 1970). In moisture

---

* Assuming a certain steady set of temperatures, humidities and building fabric characteristics; and ignoring the complex changing relationships which occur in real life situations. Such changes are or can be important with thermally lightweight constructions.

* When using the newer methods of predicting condensation, vapour pressure rather than temperature are used. This does not alter the principles described. Dewpoint for example would be described as saturated vapour pressure.

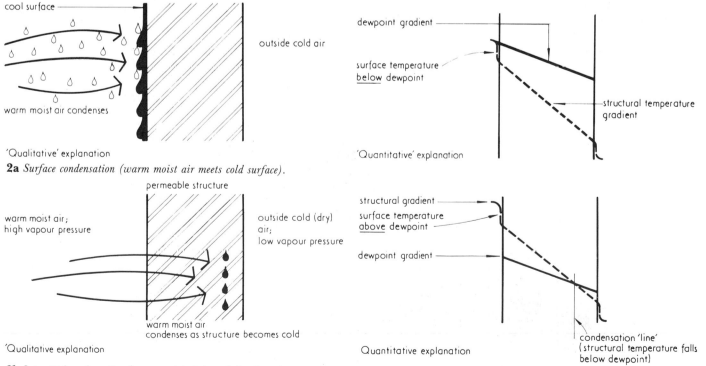

2a *Surface condensation (warm moist air meets cold surface).*

2b *Interstitial condensation (warm moist air is cooled to dewpoint temperature inside structure).*

analysis the possible drying out in the following summer of condensation within a construction during the preceding winter is considered. Depending on circumstances, it may therefore be possible to dispense with a vapour control layer (hitherto loosely referred to as a vapour barrier). Such a layer is thought by some to be better avoided, in the UK especially, to reduce the possibility of moisture entrapment during laying flat roofs. (In the UK it is seldom possible to lay a flat roof covering when there is no rain at all.)

Moisture gain analysis was first introduced tentatively in Tarmac's *Flat roofing, a guide to good practice* (1982). Later it was explained and examples given of its application in BS 6229: 1982: *British Standard Code of Practice for flat roofs with continuously supported coverings*. PSA, *Technical guide to flat roofing*, 1987, also includes worked examples of moisture gain analysis. BS5250: 1989 (revising BS 5250: 1975) has a new method of prediction. It does not use moisture gain analysis but in common with Seiffert it enables the prediction of planes along which condensation is likely. **3**. This is an important difference from older methods that predicted a zone of condensation: see page 2, caption **1**. Importantly, in the new code 'vapour control layer' (vcl) replaces both 'vapour barrier', and 'vapour check', recognising variations of vapour resistance that occur in practice. The risks involved with common forms of

construction and the ways of controlling condensation are discussed systematically and helpfully, with diagrams: see **4**. Appendices contain comprehensive hygro-thermal data and worked examples using a standard worksheet, **5**, a blank copy of which is included. (For a comparison of the current methods of prediction see 'The calculation of interstitial condensation risk' by K. A. Johnson, Building Technical File, 26.07.1989, pp. 23–30. Peter Burberry's appraisal of the new code is essential reading for a better understanding of its strengths and weaknesses: *Aj*, 26.09.90, pages 59–63).

4 Diagnostic checklist

4.01 General

Ventilation

● Is the rate of ventilation adequate to reduce the amount of moisture in the air, **6**? Ventilation itself may not necessarily prevent or reduce the risk of condensation occurring: the *rate* and *mode* of ventilation must be related to moisture emission (see below), thermal insulation and thermal response.

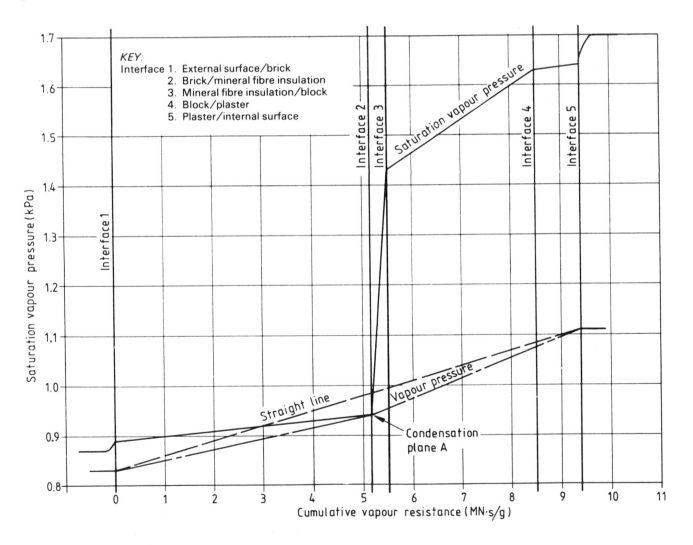

NOTE. Materials are shown in terms of their vapour resistances *not* physical thickness.

3 *Condensation is now believed to occur at interfaces between layers within a construction (described as planes in BS 5250: 1989).* (**a**), *from BS 5250: 1989, illustrates the graphical method for a wall;* (**b**), *from Burberry's*   *article, illustrates more vividly the location of planes of saturation (at which condensation occurs) in a cold deck flat roof based on an analysis using the method described in BS 5250: 1989.*

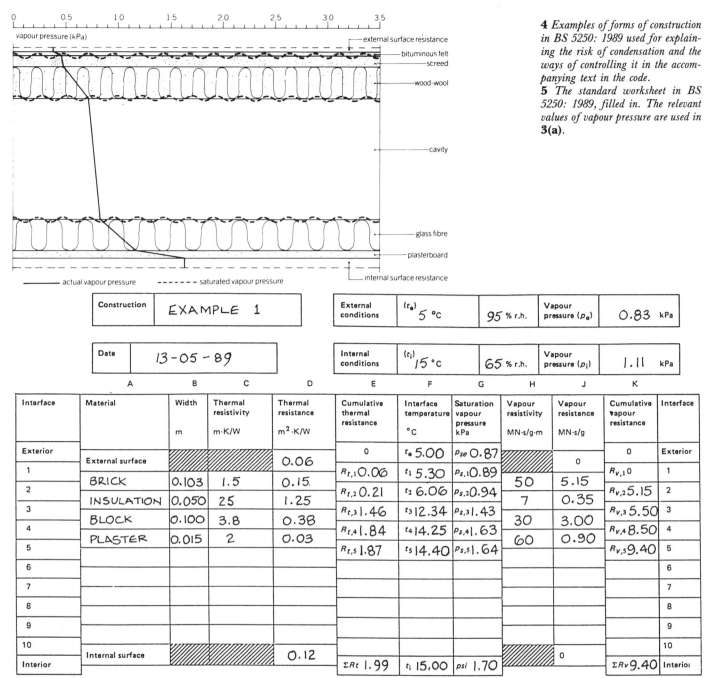

legend: —— actual vapour pressure   - - - - - saturated vapour pressure

Labels on figure: external surface resistance, bituminous felt, screed, wood-wool, cavity, glass fibre, plasterboard, internal surface resistance

vapour pressure (kPa)

4 *Examples of forms of construction in BS 5250: 1989 used for explaining the risk of condensation and the ways of controlling it in the accompanying text in the code.*

5 *The standard worksheet in BS 5250: 1989, filled in. The relevant values of vapour pressure are used in* **3(a).**

Construction	EXAMPLE 1			External conditions	(t_e) 5 °C	95 % r.h.	Vapour pressure (p_e)	0.83 kPa
Date	13-05-89			Internal conditions	(t_i) 15 °C	65 % r.h.	Vapour pressure (p_i)	1.11 kPa

	A	B	C	D	E	F	G	H	J	K	
Interface	Material	Width m	Thermal resistivity m·K/W	Thermal resistance m²·K/W	Cumulative thermal resistance	Interface temperature °C	Saturation vapour pressure kPa	Vapour resistivity MN·s/g·m	Vapour resistance MN·s/g	Cumulative vapour resistance	Interface
Exterior	External surface	///	///	0.06	0	t_e 5.00	p_{se} 0.87	///	0	0	Exterior
1					$R_{t,1}$ 0.06	t_1 5.30	$p_{s,1}$ 0.89			$R_{v,1}$ 0	1
	BRICK	0.103	1.5	0.15				50	5.15		
2					$R_{t,2}$ 0.21	t_2 6.06	$p_{s,2}$ 0.94			$R_{v,2}$ 5.15	2
	INSULATION	0.050	25	1.25				7	0.35		
3					$R_{t,3}$ 1.46	t_3 12.34	$p_{s,3}$ 1.43			$R_{v,3}$ 5.50	3
	BLOCK	0.100	3.8	0.38				30	3.00		
4					$R_{t,4}$ 1.84	t_4 14.25	$p_{s,4}$ 1.63			$R_{v,4}$ 8.50	4
	PLASTER	0.015	2	0.03				60	0.90		
5					$R_{t,5}$ 1.87	t_5 14.40	$p_{s,5}$ 1.64			$R_{v,5}$ 9.40	5
6											6
7											7
8											8
9											9
10											10
	Internal surface	///	///	0.12				///	0		
Interior					ΣR_t 1.99	t_i 15.00	psi 1.70			ΣR_v 9.40	Interior

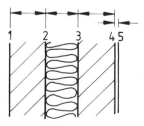

(a) Full cavity fill

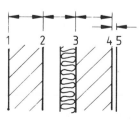

(b) Partial cavity fill

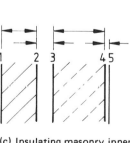

(c) Insulating masonry inner leaf

1 to 2 Masonry
2 to 3 Full cavity fill or airspace plus partial cavity fill or airspace
3 to 4 Masonry
4 to 5 Internal finish

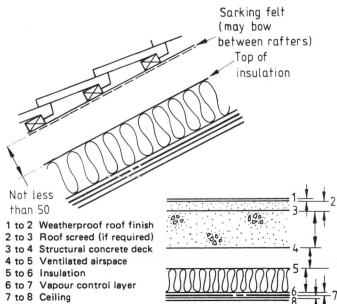

Sarking felt (may bow between rafters)
Top of insulation
Not less than 50

1 to 2 Weatherproof roof finish
2 to 3 Roof screed (if required)
3 to 4 Structural concrete deck
4 to 5 Ventilated airspace
5 to 6 Insulation
6 to 7 Vapour control layer
7 to 8 Ceiling

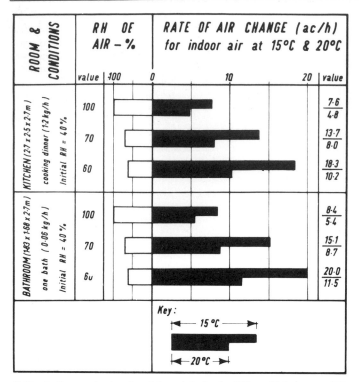

6 *Facsimile reproduction from* Materials for building, *Vol. 4, page 129, showing relationship between air change rates and relative humidities for different conditions of moisture emission and air temperature. The relative humidities are those to be maintained; the air change rates are those necessary to maintain the relative humidities.*

Moisture emission
- The amount of moisture produced within the building or spaces within the building can be considerable. Check sources below, and decide whether occupancy conditions are more severe than average.
- In dwellings, sources of moisture emission include: cooking and dishwashing; floor cleaning; clothes washing and drying; heating (paraffin and flueless gas heaters); bathing (including showering); plants and breathing (notably in bedrooms at night). See **7, 8**.

SOURCE OF MOISTURE		PERIOD OF EMISSION	MOISTURE EMISSION — kg (for period stated) value			
			value 0	5	10	15
KITCHEN	Normal actual (1)	day	3·70			
	Cooking & dishwashing (2)	day	2·57			
	Gas cooking (3)	h	0·64			
	" " (3 meals) (1)	day	2·10			
	Kettle boiling (3)	h	1·65			
	Dishwashing (3 meals) (1)	day	0·50			
	Floor mopping/m² (2)	each wash	0·15			
	Sink full of hot water (3)	h	0·50			
Clothes	washing (1)	day	2·00			
	drying (1)	day	12·00			
3kw washing machine (3)		h	3·85			
Paraffin heater (4)		h	0·35			
Plant (watering & respiration)		day	0·84			
Shower (2)		each	0·23			
Bath (2)		each	0·05			
Adult	at rest or asleep (4)	day	0·40			
	sweat (av.) (3)	day	0·70			

7 *Facsimile reproduction from* Materials for building, *Vol. 4, page 127, showing comparison of average rates of water emission from various sources.*

Table 1. Typical moisture generation rates for household activities

Household activity	Moisture generation rate
People:	
asleep	40 g/h per person
active	55 g/h per person
Cooking:	
electricity	2000 g/day
gas	3000 g/day
Dishwashing	400 g/day
Bathing/washing	200 g/person per day
Washing clothes	500 g/day
Drying clothes indoor (e.g. using unvented tumble drier)	1500 g/person per day

8 *Facsimile reproduction from BS 5250: 1989, of table allowing room air moisture contents to be estimated for common domestic situations.*

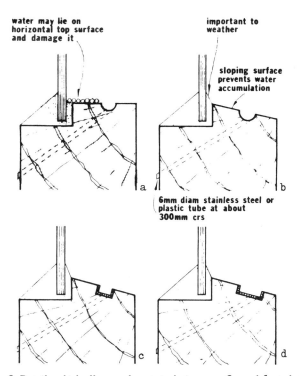

9 *Details which allow condensate to drain away.* **6c** *and* **d** *are better than* **b** *(but* **c** *may be difficult to clean).* **6a** *does not allow drainage and should be avoided.*

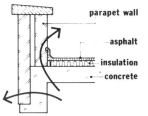

10a *An example of cold bridging (see also overleaf).*

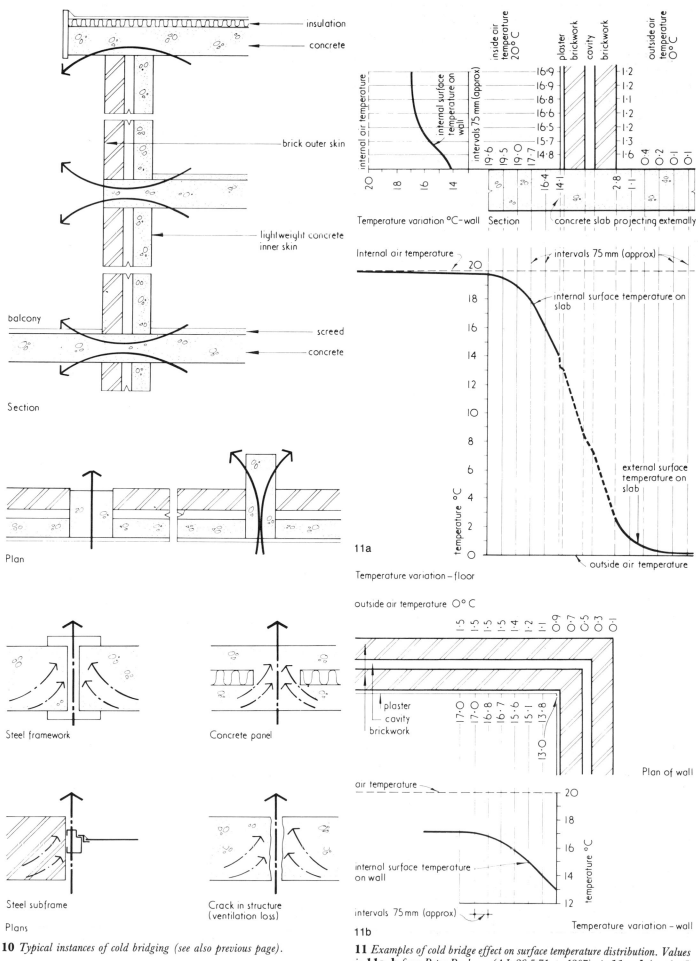

Section

Plan

Steel framework

Concrete panel

Steel subframe

Crack in structure
(ventilation loss)

Plans

10 *Typical instances of cold bridging (see also previous page).*

Temperature variation °C – wall   Section   concrete slab projecting externally

11a   Temperature variation – floor

11b   Plan of wall / Temperature variation – wall

11 *Examples of cold bridge effect on surface temperature distribution. Values in* **11a, b** *from Peter Burberry (AJ, 26.5.71, p. 1207); in* **11c, d** *(overleaf) from E. F. Ball, BRS Current Paper (Design series) 24, Fig. 8.*

outside air temperature –1·1° C

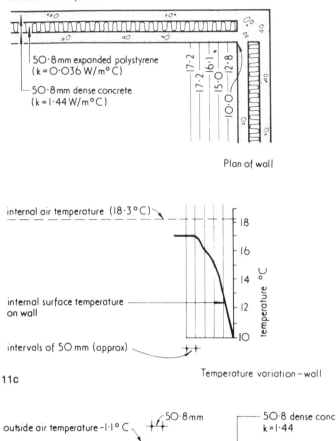

50·8mm expanded polystyrene
(k = O·O36 W/m° C)

50·8mm dense concrete
(k = 1·44 W/m° C)

17·2 16·1 12·8
17·2 15·O 1O·O

Plan of wall

internal air temperature (18·3°C)

internal surface temperature
on wall

intervals of 50 mm (approx)

temperature °C

Temperature variation – wall

11c

outside air temperature –1·1° C — 50·8mm

50·8 dense conc
k = 1·44

50·8mm expanded polystyrene
k = O·O35

inside air temperature 18·3° C

Plan
50·8mm bridge

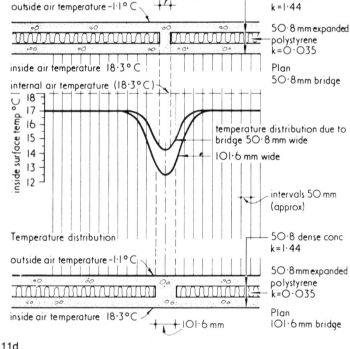

internal air temperature (18·3°C)

inside surface temp °C

temperature distribution due to
bridge 50·8 mm wide

1O1·6 mm wide

intervals 50 mm
(approx)

Temperature distribution

50·8 dense conc
k = 1·44

outside air temperature –1·1° C

50·8mm expanded polystyrene
k = O·O35

inside air temperature 18·3° C

1O1·6 mm

Plan
1O1·6 mm bridge

11d

Mode of heating
- Continuous or intermittent mode; and relationship with thermal response of the fabric. Continuous heating better suited to slow-response buildings; intermittent heating better suited to quick-response ones.
- Condensation is more likely when the air temperature is raised quickly and the surface temperature rises slowly (i.e. slow-response fabric).

lintel   jamb   sill

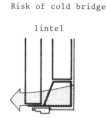

Avoiding action

1 Continue insulation to back of frame

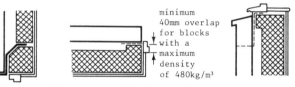

minimum
40mm overlap
for blocks
with a
maximum
density
of 480kg/m³

2 Overlap window frame and insulating blockwork

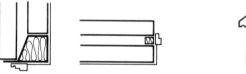

3 Fill lintel with insulation and use insulating cavity closers

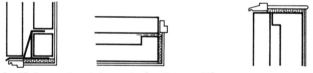

4 If frame placed forward, insulate soffit, reveal and under sill

12 *Redrawn diagram from BRE Report, Thermal insulation: avoiding risks, 1989, page 17. Building Research Establishment, Crown Copyright.*

Removal of condensate
- If allowed to occur, is condensate removed regularly or drained away, **9**? (Small amounts of condensate are unlikely to cause other problems.)

4.02 Surface condensation

Cold bridges
- Corners, junctions or situations where materials of high conductivity are used in a construction (e.g. concrete column in cavity wall with insulating inner leaf), **10**.
- Cold bridge effects may be responsible for persistent condensation in localized areas, **11**. In some locations in the vicinity of cavity bridging situations (window surrounds, for instance), dampness may be due to rainwater penetration rather than condensation.

Surface temperature/insulation relationship
- For any given indoor/outdoor temperature difference and internal surface resistance, the internal surface temperature *increases* as the U-value (thermal transmittance) *decreases*, and surface condensation is less likely, **12**.

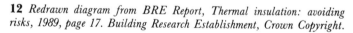

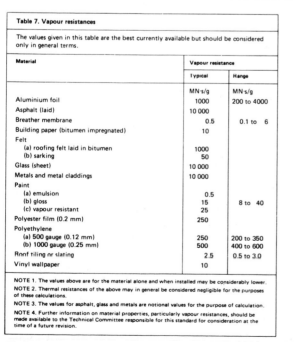

Table 6. Thermal and vapour resistivities

The values given in this table are the best currently available but should be considered only in general terms.

Material	Thermal resistivity	Vapour resistivity	
		Typical	Range
	m·K/W	MN·s/g·m	MN·s/g·m
Airspace	See table 8	5	
Asbestos cement sheeting and substitutes	2.5	300	200 to 1000
Asphalt (laid)	2.0	See table 7	
Blockwork			
(a) lightweight (800 kg/m³)	3.8	30	20 to 50
(b) medium weight (1400 kg/m³)	1.8	50	30 to 80
(c) dense (2000 kg/m³)	0.8	100	60 to 150
Brickwork			
(a) common/facing (1500 kg/m³)	1.5	50	25 to 100
(b) sandlime (1500 kg/m³)	1.5	100	80 to 200
(c) engineering (2000 kg/m³)	0.8	120	100 to 250
Carpeting			
(a) normal backing	20.0	10	7 to 20
(b) foam backed or with foam underlay	10.0	200	100 to 300
Chipboard	7.0	500	
Concrete (cast)			
(a) lightweight (1800 kg/m³)	1.0	40	30 to 80
(b) dense (2200 kg/m³)	0.8	200	
(c) no fines (1800 kg/m³)	1.0	20	
Cork board	24.0	100	50 to 200
Fibreboard (sheet or ceiling tile)	15.0	40	15 to 60
Fibre (glass or rock)	25.0	7	
Glass			
(a) sheet	1.0	See table 7	
(b) expanded or foamed	16.0	10 000	
Hardboard	8	600	450 to 1000
Metals or metal claddings	0.02	See table 7	
Phenolic (foamed (closed cell))	50.0	300	200 to 750
Plaster	2.0	60	
Plasterboard	6.0	60	
Polystyrene			
(a) expanded bead	30.0	300	100 to 600
(b) expanded extruded	40.0	1000	600 to 1300
Polyurethane (foamed (closed cell))	45.0	600	500 to 1000
Plywood			
(a) sheathing	7	450	150 to 1000
(b) decking	7	2000	1000 to 6000
PVC (polyvinyl chloride) sheet or tile	1.2	1000	800 to 1300
Rendering	1.8	100	
Roof (tiling or slating)	1.2	See table 7	See table 7
Roofing felt	2.0	See table 7	See table 7
Screed			
(a) aerated	2.5	100	
(b) cast	0.8	200	
Stonework			
(a) granite, slate and marble	0.5	300	150 to 450
(b) limestone and sandstone	0.5	200	150 to 450
Tiling (ceramic)	0.6	2000	500 to 5000
Timber	7	60	40 to 70
Urea formaldehyde	25	15	10 to 30
Vermiculite	15	15	
Woodwool slabs	10	20	15 to 40

NOTE. Further information on material properties, particularly vapour resistivities, should be made available to the Technical Committee responsible for this standard for consideration at the time of a future revision.

Table 7. Vapour resistances

The values given in this table are the best currently available but should be considered only in general terms.

Material	Vapour resistance	
	Typical	Range
	MN·s/g	MN·s/g
Aluminium foil	1000	200 to 4000
Asphalt (laid)	10 000	
Breather membrane	0.5	0.1 to 6
Building paper (bitumen impregnated)	10	
Felt		
(a) roofing felt laid in bitumen	1000	
(b) sarking	50	
Glass (sheet)	10 000	
Metals and metal claddings	10 000	
Paint		
(a) emulsion	0.5	
(b) gloss	15	8 to 40
(c) vapour resistant	25	
Polyester film (0.2 mm)	250	
Polyethylene		
(a) 500 gauge (0.12 mm)	250	200 to 350
(b) 1000 gauge (0.25 mm)	500	400 to 600
Roof tiling or slating	2.5	0.5 to 3.0
Vinyl wallpaper	10	

NOTE 1. The values above are for the material alone and when installed may be considerably lower.

NOTE 2. Thermal resistances of the above may in general be considered negligible for the purposes of these calculations.

NOTE 3. The values for asphalt, glass and metals are notional values for the purpose of calculation.

NOTE 4. Further information on material properties, particularly vapour resistances, should be made available to the Technical Committee responsible for this standard for consideration at the time of a future revision.

13 *Facsimile reproduction from BS 5250, of tables giving vapour resistance of membranes, and vapour resistivity of materials. Building Research Establishment, Crown Copyright.*

- A more or less even distribution of condensation on the inner face of a wall or roof slab would suggest a poorly insulated fabric (i.e. value too high) for the prevailing climatic conditions, indoors and outdoors.

Thermal response of the fabric
- Condensation is likely to occur more frequently and at a faster rate in buildings with thermally heavyweight (i.e. slow-responding) fabrics if they are intermittently heated: the air is heated quickly when the heating system is turned on, but the fabric surface remains cold for some time, thereby inviting surface condensation.
- The rapidity of response of heavyweight fabrics can be made significantly faster by adding low-conductivity materials to the inner face (insulated linings are notable), but this may lead to interstitial condensation if no effective vapour control layer is provided on the warm side of the insulation layer.

4.03 Interstitial condensation

Vapour control layers
- Adequacy of material used to provide the requisite vapour resistance must be checked, **13**.
- Continuity of material used is essential, to ensure complete overall vapour resistance.
- Temperature at surface of material to be used must be checked. To be effective it must be *above* dewpoint.

Walls
- Check resistance of vapour contol layer as above.
- Can condensate drain away harmlessly to the outside?
- Check for the unwitting incorporation of a material externally or within the thickness of a construction that could act as a vapour control layer but in the wrong position (vapour control layers should always be on the warm side of insulating materials). Could be metal, plastic or porous material with an impervious coating on the external (cold) face.

Flat roofs
(a) Timber
- Check resistance of vapour control layer as above. (Are there any? Are they correctly positioned? Are they effective?)
- If a *cold* roof (insulation at ceiling level), is the ventilation within the roof space adequate? (Ventilation openings should be on opposite sides of the roof, and should be equivalent in area to at least 0.4 per cent of the total plan area or 0.6 per cent if the building is sheltered.)
- If a *warm* roof (insulation at roof finish level), has the roof void been properly sealed to ensure that there is no ventilation?

(b) Concrete
- Check resistance of vapour control layer as under Timber, above.
- Check that ventilation for allowing escape of entrapped moisture is not also ventilating a warm roof (see Study 21: Entrapped moisture).

Pitched roofs
- Is the underlay under tiles, slates, etc. acting as a vapour control layer? If so (and particularly if the ceiling is insulated) is the roof space adequately ventilated? It must be.

Note: Saturated bitumen or polythene felts have high vapour resistance and can act as unintended vapour control layers preventing escape of moisture laden air. Special 'breather' felts are available.

5 Further reading

1 Addleson, Lyall, *Materials for building*, Vol. 4, London, Newnes-Butterworths, 1975, 4.06 Condensation, pages 124–142.
Methods of prediction now overtaken by techniques based on moisture gain analysis, otherwise comprehensive and detailed coverage of principles with nomograms and charts.

2 BS 5250: 1989: *Code of practice for control of condensation in buildings*. A major revision of BS 5250: 1975, retitled to include all buildings. Many more pages (120 compared with 27); design principles applied to specific constructional examples with related risks and how to control condensation; introduces a different method for predicting condensation risk at planes(s) within a construction using the standard worksheet provided; and, importantly prefers 'vapour control layer' to either 'vapour barrier' or 'vapour check'. (See reference 13 for analysis of the new code and its relationship with existing methods of prediction and reference 14 for an appraisal). Extensive guidance on diagnosis and remedial work in section four, pages 44ff. (MPBW reference 4, below, though dated is still useful for principles). Recommendations for advice to building owners/occupiers is given in section six, pages 49 and 50. This is succinct and clear.

3 BS 6229: 1983: *British Standard Code of Practice for flat roofs with continuously supported coverings.* includes concept of moisture gain analysis and method of its prediction.

4 Burberry, Peter, *Condensation and how to avoid it. AJ Energy File*, 3 October 1979, pages 723–739.
A thorough study which covers the physical principles of condensation, case studies of its occurrence, how to predict its risk in new work and how to control it in existing buildings. Does not include moisture gain analysis.

5 DOE, *Condensation in dwellings*, Parts 1 and 2, London, HMSO, 1970 and 1972.
The simplest, most concise and explicit explanation of condensation, very well illustrated (in colour) and with general guidance on both design precautions and remedial work. Some parts are now a little outdated by BS 5250: 1975 and BS 6229: 1982, but Part 2 gives good practical advice on investigations and remedial work. Part 1 is now out of print.

6 BRE Digests.
110 *Condensation* (new edition, 1972).
Useful for principles only.
139: *Control of lichens, moulds and similar growths*, 1977.
Lists the kinds of growths found and where, and recommends a range of toxic washes for killing them off.
180 *Condensation in roofs*, August 1975.
Discusses the design principles for minimising condensation risk and includes latest thinking (incorporated into BS 5250: 1975) on fundamental difference between cold and warm roofs. Methods of prediction not included.
297 *Surface condensation and mould growth in traditionally-built dwellings.* Useful supplement to Digest 139.

7 BRE Defect Action Sheets
1 *Slated or tiled pitched roofs: ventilation to outside air* (Design) gives essential basic guidance.
3 *Slated or tiled pitched roofs: restricting the entry of water vapour from the house* (Design) explains methods of effecting a proper seal.
4 *Pitched roofs: thermal insulation near eaves* (Site) explains how to achieve continuity of ventilation of the roof void.
14 *Wood windows: preventing decay* (Design) for advice on drainage grooves
16 *Walls and celings: remedying recurrent mould growth* is a good summary.
59 *Felted cold deck flat roofs: remedying condensation by converting to warm deck* is a succinct text, good on precautions. Helpful three-dimensional illustrations.
77 *Cavity external walls: cold bridges around windows and doors* (Design).
78 *Cavity external walls – dry lining: avoiding cold bridges* (Design).

8 BRE Information papers
11/85, *Mould and its control.*
13/87, *Ventilating cold deck flat roofs.* Explains why there may be a need to increase current minimum aperture of 0.4 per cent of the plan roof area (BS 6229) to 0.6 per cent for complex plan shapes or for simple plan shapes of low buildings or buildings located in a sheltered location.

9 BRAS Technical Information Leaflets
59 *Condensation in domestic tiled pitched roofs – advice to householders.*

10 BRE Report *Thermal insulation: avoiding risks*, 1989. Comprehensive guide explaining the technical risk which may be assoicated with meeting the building regulation requirements for thermal insulation. Well illustrated with relevant details for roofs, walls, windows and floors and explains what actions could be taken to avoid the risk of failure. Has many illustrations confirming the need for continuity of insulation and vapour control layers, all appropriate to underline the principle of continuity – see Principles for building page 3.

11 Tarmac's *Flat roofing, a guide to good practice*, 1.2 Thermal design and 1.3 Vapour design guide. First to introduce in book form the concept of moisture gain analysis, by which is may not be necessary always to have a vapour control layer in a warm deck flat roof design. No methods of prediction nor properties are given but it has useful tables and a summary covering a wide range of constructions and internal temperature conditions.

12 *AJ* Series, *Construction Risks and Remedies, Condensation*, 9.4.86, pages 49–58, and 16.5.86, pages 69–81.
Discusses the basic mechanics of condensation, the causes of the most important problems and how they can be avoided or resolved.

13 Pilkington Research and Development, The calculation of interstitial condensation risk by K. A. Johnson, Building Technical File, 26, July 1989, pages 23–30.

14 Burberry, Peter, *Controlling the risk, Code for condensation*, AJ Technical, 26.09.90, pp. 59–63.

Study 21
Entrapped moisture

1 Location of failures

1.01 flat roofs

This is the most common location of entrapped moisture, and the effects include:

- loss of thermal insulation, which in turn increases the incidence of condensation and mould growth
- decay and deterioration of organic and moisture-sensitive materials (e.g. timber and timber products;

1 *Leakage of entrapped moisture from a discontinuity in the roof slab.*

mild steel reinforcement in concrete; ferrous and metal fixings)
- staining of ceilings, through leaching of bitumen from the waterproof roof covering by alkalis derived from the cement in concrete slabs and/or screeds, **1**
- blistering and subsequent tearing of the waterproof roof covering, through vaporisation of moisture underneath the covering during exposure to sunlight, **2**.

2 Definition

The retention of moisture, usually as water vapour, within the thickness of a construction (generally a roof), the externally exposed surface of which is covered continuously with a waterproof membrane.

3 Sources of moisture

3.01 During construction

- Construction water, i.e. water used in mixing concrete (for the slab) or in mixing cement-based screeds. Lightweight screeds usually require large amounts of mixing water (for instance, a lightweight aerated cement screed 75 mm thick may require up to $11 \, l/m^2$ of mixing water).
- Absorbed rainwater, that could equal or even exceed the amount of construction water. For example, the measured absorption of a screed 75 mm thick with a fall of 1:60 was about $8 \, l/m^2$ after four days of rain.

The same applies to rainwater falling on dry insulants. Entrapment may be aggravated if there is a vapour control layer under the insulant, as that layer prevents water

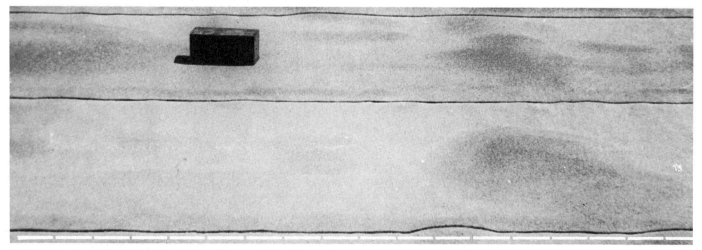

2 *Blistering of roof felt. Unlike, **5**, raised profiles are in island form rather than continuous corrugations; but causes are similar.*

penetration through it. (The dual function of a vapour control layer – i.e. controlling water vapour flow and excluding liquid water – should also be remembered.)

3.02 After construction
Water produced by condensation within the thickness of the construction, in particular under the waterproof roof covering, for instance as a result of *interstitial* condensation (see Study 20: Condensation, 4.03, page 108). The risk of such condensation occurring depends on the interrelationship of:

- the moisture content of the air contiguous with the internally and externally exposed surfaces
- the temperature gradient through the construction (thermal resistance of layers of material within the construction are relevant)
- the dewpoint gradient through the construction (vapour resistances of layers of material are relevant).

All of these points are discussed at greater length in Study 20: Condensation, 4.03.

Note: Interstitial condensation alone, rather than rainwater or construction water, may be responsible for the amount of moisture entrapped; or, more commonly, it may increase the amount of moisture already retained during construction. It is therefore usually extremely difficult to identify precisely the main source of moisture, though the diagnostic checklist in this study and Study 20: Condensation, page 106 will help.

4 Basic mechanisms

The phases associated with entrapped moisture are:

4.01 Entrapment
Moisture becomes entrapped mainly because it has only *one* surface (the inside surface) from which to evaporate, the outer surface being sealed by the waterproof membrane.

Transmission of moisture (as water vapour) from the interior of the construction to this evaporating surface is difficult and, apart from being affected by the pore structure of the materials within the construction, it is influenced also by thermal effects, temperature gradients across the construction in particular.

Evaporation from the inside surface is generally slow, unless there are good drying-out conditions, i.e. adequate heat and ventilation (see Study 24; Movement, 8.03, page 143).

4.02 Moisture transmission
Moisture trapped within a construction will, at first, be generally evenly distributed, i.e. be in a condition of equilibrium. Later the equilibrium is broken and the moisture moves, subsequently gravitating towards and settling in ponds or hollows of the roof slab or at the bottom of falls, often near rainwater outlets.

Further movement of moisture takes place as a result of temperature gradients, the moisture tending to migrate from warmer to cooler zones. Seasonal variations in temperature tend to cause the lower part of the construction to be wetter than the upper in summer (good for evaporation) with the converse occurring during the winter (good for condensation under the waterproof roof covering), **3**.

4.03 Condensation
Water vapour (from the moisture already entrapped, or from the air from the building interior, or both) reaching the underside of a cold (in winter) waterproof roof covering will condense on the underside. The condensed water may at first be distributed evenly within the construction, and then find its way back to the underside of the roof construction where it may evaporate (see 4.02, 'Moisture transmission', above). Alternatively and perhaps more commonly, the equilibrium condition is upset by cracks and other discontinuities in the construction (e.g. electrical conduits) and the condensed water will instead drain at localized points from which it may also drip, **4**.

4.04 Vaporisation
Solar radiation on the roof in summer will cause the entrapped moisture within the construction to vaporise.

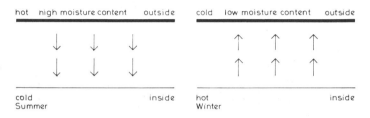

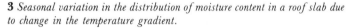

3 *Seasonal variation in the distribution of moisture content in a roof slab due to change in the temperature gradient.*

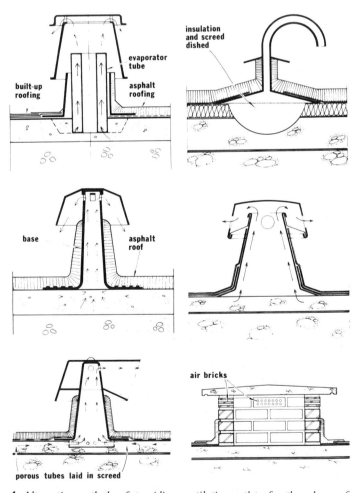

4 *Alternative methods of providing ventilation outlets for the release of entrapped moisture (insulation not shown). Manufacturers should be consulted for details. These include D. Anderson & Son Ltd; Screeduct Ltd; Briggs Amasco Ltd; Langley London Ltd and Aluminium Developments. Cheecolite Ltd produce a cavity heat-dried screed.*

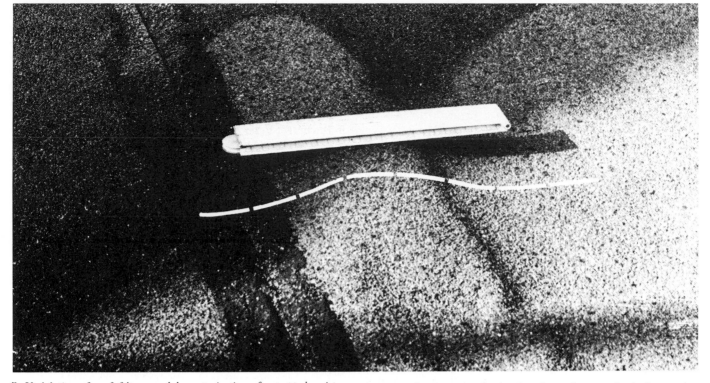

5 *Undulation of roof felt, caused by vaporisation of entrapped moisture and/or insufficient or badly distributed bitumen compound. Raised profile will yield to pressure, indicating that undulation is not in underlying structure.*

Considerable pressure is exerted on the underside of the waterproof roof covering by this water vapour.

The pressure will:

- *at best*, weaken adhesion of felt roofs (the adhesive is in any case already weakened by the heat from the radiation), or
- *at worst*, cause small areas of the covering (felt or asphalt) to blister, **2**, or to tear.

Vapour pressure is relieved by the formation of the blisters. But the blisters may be subsequently torn (thus allowing ingress of rainwater) by foot traffic or by frost action, if the blisters become filled with water.

4.05 Cross reference
See Studies 5 and 6 for further information on failures of the above types, in the structural contexts in which they occur (i.e. waterproof membranes and flat roof slabs). See, in particular, Study 6, Manifestations 5 and 6, pages 35–37.

5 Diagnostic checklist

5.01 Reduction of moisture content
Were steps taken to reduce the moisture content in the roof construction before the waterproof covering was laid? This could have been done by the following methods.

- By inserting tubes in or drilling holes through the roof slab, thus helping to drain away excess water. Were these provided?
- Weather protection may be essential to avoid dry insulants in particular absorbing rainwater. Was such protection provided?
- The choice of an alternative screed that uses little water (e.g. no fine lightweight screed with cement:sand blinding, mortar topping) or no water at all (specially

treated aggregates) may be the only practical alternative to conventional temporary covering lifted off the deck to allow ventilation. Was this done?

5.02 Reduction of vapour pressure
Were special ventilators installed to reduce the vapour pressure under the waterproof covering? Specially bent copper tubes and what are known as 'breather vents' may be installed to reduce the vapour pressure under the waterproof roof covering but they do not normally enable much of the entrapped moisture as such to evaporate. To be effective, the vents need to be well distributed over the area of the roof, **4**.

5.03 Drying out
Was provision made for the construction to dry out after completion of the building?

Effective drying out of the construction after completion demands the provision of adequate ventilation underneath the waterproof roof covering. To achieve the ventilation required in turn demands the provision of airways in the form of small ducts within the screed and adequate air inlets and outlets (the area of edge ventilators may have to vary between 1100 mm and 4300 mm^2/m run); rooftop ventilators may have to be spaced as close as 6 m apart and located at both low and high points of the roof.

Whatever system of ducting and ventilators is used, it is unlikely that screeds will dry out much before one year has elapsed. The process may be accelerated by using low-speed fans in conventional roof ventilators connected to the roof ducting. The speed of drying out needs to be controlled so as to reduce the risk of damage to the screed (Study 24: Movement).

Theoretically, ventilators may be removed once drying out has been completed, but in practice it is usually advisable to maintain some ventilation so as to reduce the risk of interstitial condensation – but see 'Note' below.

Note
There is a real conflict during the winter between ventilation to assist drying out and ventilation that may

increase the risk of interstitial condensation and/or reduce thermal insulation. It would now appear that ventilation should only be provided within the roof construction if the thermal insulation is located *below* the ventilation ducts or space, i.e. a cold deck construction. This usually means insulation placed on or near the soffit of the roof deck.

6 Further reading

1 Addleson, Lyall and Rice, Colin, *Performance of materials in buildings*, Butterworth-Heinemann, 1992, 3.2 Exposure, Entrapped moisture, pages 214–222. Detailed, updated and expanded coverage of mechanisms and precautions in Addleson, Lyall, *Materials for building*, Vol. 2, London, Iliffe Books, 1972, pages 156–161.

2 BS 5250: 1989: *Code of practice for control of condensation in buildings*. A major revision of BS 5250: 1975 and retitled to include all buildings. Many more pages (120 compared with 27); design principles applied to specific constructional examples with related risks and how to control condensation; introduces a different method for predicting the risk of condensation (in terms of its rate over a 60 day period) at plane(s) within a construction using the standard worksheet provided; and, importantly prefers 'vapour control layer' to either 'vapour barrier' or 'vapour check'. (See reference 13 for analysis of the new code and its relationship with existing methods of prediction and reference 14 for an appraisal of the new code). Extensive guidance on diagnosis and remedial work in section four, pages 44ff. (MPBW reference 4, below, though dated is still useful for principles). Recommendations for advice to building owners/occupiers is given in section six, pages 49 and 50. This is succinct and clear.

3 BS 6229: 1983: *British Standard Code of Practice for flat roofs with continuously supported coverings* includes concept of moisture gain analysis and method of its prediction.

4 Tarmac's *Flat roofing, a guide to good practice*, 1.2 Thermal design and 1.3 Vapour design guide. First to introduce in book form the concept of moisture gain analysis, by which it may not be necessary always to have a vapour control layer in a warm deck flat roof design. No methods of prediction nor properties given but has useful tables and a summary covering a wide range of constructions and internal temperature conditions.

5 Burberry, Peter *Condensation and how to avoid it. AJ* Energy File, 3 October 1979, pages 723–739.
A thorough study which covers the physical principles of condensation, case studies of its occurrence, how to predict its risk in new work and how to control it in existing buildings.

6 DOE, *Condensation in dwellings* Parts 1 and 2, London, HMSO, 1970 and 1972.
The simplest, most concise and explicit explanation of condensation, very well illustrated (in colour) and with general guidance on both design precautions and remedial work. Some parts are now a little out-dated by BS 5250: 1975 but Part 2 gives good practical advice on investigations and remedial work. Part 1 now out of print.

7 BRE Digests:
110, *Condensation* (new edition, 1972) – for principles.
Useful for methods of prediction.
163, *Drying out buildings* (1974).
Describes ways of drying out, including dehumidifiers and methods of testing the moisture content of walls, floors and joinery.
180, *Condensation in roofs* (August, 1975).
Discusses the design principles for minimising condensation risk and includes latest thinking (incorporated into BS 5250: 1975) on fundamental differences between cold and warm roofs. Methods of prediction not included.

8 Burberry, Peter, *Controlling the risk, Code for condensation*, AJ Technical, 26.09.90, pp. 59–63.

Study 22
Rain penetration

1 Location of failures

The following are the most common locations of failures.

- Flat roofs (waterproof membrane, including skirtings; parapets). See Studies 6 and 9, pages 29 and 48.
- Walls (cavity construction in bricks and blocks; or walls in solid concrete). See Study 13, page 66.
- Around windows. See Study 16, page 85.
- Joints in claddings (concrete panels; thin metal or glass panels). See Studies 14, 15, 17 pages 78, 83 and 92.

2 Effects of rain penetration

Rain penetration has the following effects which underlie failure, an awareness of which should precede diagnosis.

2.01 Loss of adhesion and/or cracking of finishes
These are usually due to:

- chemical reaction with constituents of the background material (for instance, in efflorescence or sulphate attack)
- frost action
- differential movement of finish or background.

2.02 Cracking of components or elements
This is usually due to:

- moisture movement during wetting or drying out
- chemical reactions (e.g. corrosion of mild steel reinforcement in concrete or non-ferrous metal fixings in concrete or masonry).

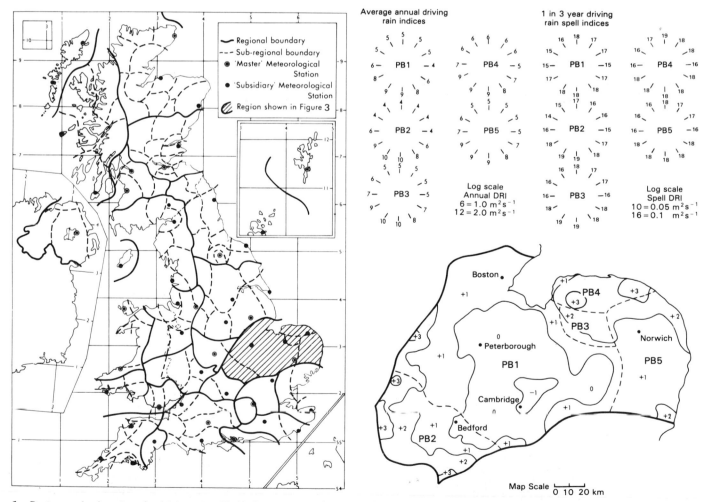

1a *Regions and sub-regions for driving rain. (Left) the regions and (right) sub-regions with rain indices. Simplified version of maps in DD 93. Facsimile reproduction from Prior.*

1b *(facing) Annual driving rain indices for sub-regions for spell (upper maps) and annual (lower maps): lowest values for south-facing walls (left) and north-facing walls (right). Facsimile reproduction from Prior.*

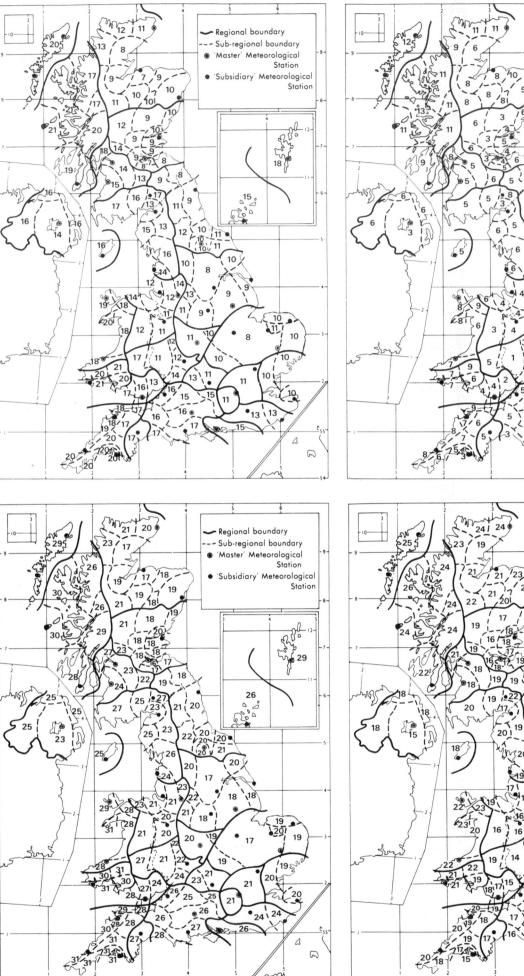

2.03 Increased condensation and mould growth
This is usually due to a reduction of the thermal resistance of some or all of the layers within the construction.

2.04 Change of appearance of surface
Such as streaking, staining, mould growth.

3 Causes of rain penetration

3.01 Inadequate impermeability of porous materials
This can apply to a number of elements but mainly affects walls. Porous materials, such as fireclay, calcium silicate or lightweight concrete are not normally expected to be completely impermeable and do absorb some rainwater, which can cause future trouble.

The permeability of walls is also dependent on the mortar jointing, but rain penetration is more usually through cracks between the mortar and units, or sometimes through cracks in the mortar joint rather than through the mortar itself.

3.02 Cracking of units of material or between components
This is due mainly to movements; see Study 24 page 134. Such cracking is liable to occur in:

- monolithic concrete walls (at day joints especially)
- waterproof membranes to flat roofs, including cracking or detachment of the skirting
- external rendered finishes
- mortar jointing, including loss of adhesion at joint/unit interface.

3.03 Inadequate provision for drainage
This applies to the third line of defence (adequate impermeability and prevention of cracks being the first and second). For example: damp-proof courses, cavity trays and flashings in open drained joints.

4 Contributory factors to rain penetration

4.01 Main characteristics
The extent to which rainwater may be prevented from penetrating into or through the construction of an element depends on the combined effects of:

1 the direction of the plane of the element and its exposure to rainfall (wind-driven rain in particular)
2 the water-excluding characteristics of the units, components and joints making up the construction of the element, and
3 the modes of transmission of the water within the thickness of the construction of the element.

4.02 Exposure
- Whatever the amount of rain, *horizontal* surfaces are basically more severely exposed as they receive more rainwater than do vertical ones; and *impervious* vertical surfaces will have a greater run-off of rain water (thus imposing greater demands on joints, etc.) than will absorptive vertical surfaces.
- Whatever the basic exposure of a site may be, *wind-driven* rain causes the most severe conditions of exposure.
- Some walls of a building will be more severely exposed than others, depending primarily on the direction of the prevailing wind-driven rain (see diagrams **1** and **3**).

2 *The basic exposure category of a site may need to be modified to take account of site topography or height of buildings as shown below.*

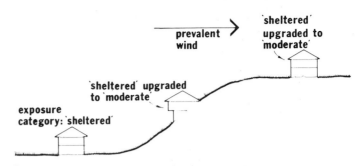

2a *Basuc 'sheltered' category modified by topography of site.*

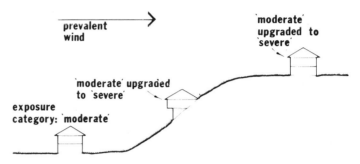

2b *Basic 'moderate' category modified by typography of site.*

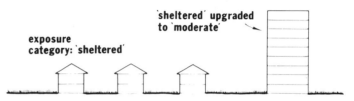

2c *Basic 'sheltered' category modified by building height.*

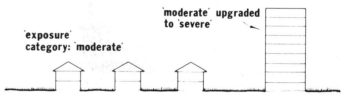

2d *Basic 'moderate' category modified by building height.*

- The severity of the basic exposure of a building to wind-driven rain will be modified (i.e. increased or decreased) by:
  (a) the nature of the topography surrounding the building (e.g. high ground, trees and other buildings – see **2**)
  (b) the geometry of the building itself and of its details.

Note on DD 93: 1944 Draft for Development Methods for assessing exposure to wind-driven rain
Using more comprehensive metereological data and computer analyses of it the Metereological Office has been able to produce more realistic values of the exposure of walls to prolonged wind-driven rain than could be obtained from the BRE driving-rain index (DRI). Whereas the BRE all-direction annual DRI was calculated on annual data, the new index is calculated from hourly data. DD 93 gives two methods of assessing exposure: the first gives the *annual index* and relates to the BRE annual index; the second the *spell index*, a new index in which account is taken of the

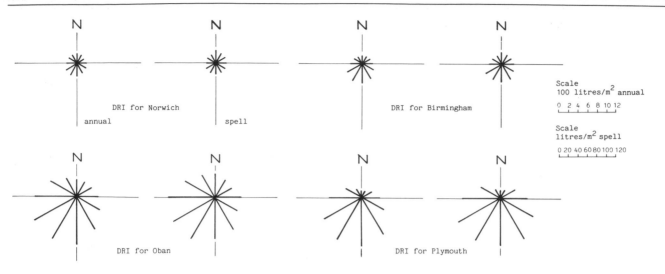

3 *Driving-rain indices as roses (after Lacy), annual and spell, for selective locations. Values are from DD 93. Note that the scales for annual and spell are not the same.*

amount of wind-driven rain over a short period (3 days) to be expected falls over the worst spell (once in three years). Both indices are expressed in l/m^2 (instead of m^2/s in the BRE DRI) and can be predicted for any of twelve wind directions. The annual index is for use when considering the average moisture content of exposed building materials for assessment of their durability. The spell index is for use when assessing the resistance of a wall to water penetration or when designing windows.

New maps, **1**, have been prepared. These are markedly different from those for the BRE DRI and include sub-regions each of which have related spell and annual rose values for 12 different wall orientations. (Comparative 'roses' after Lacy are illustrated in **3**.) In DD 93 there are standard worksheets for the computation of the spell and annual indices together with rules for modifying the exposure to take into account the effects of protection and local ground height. Completing the worksheets is tedious rather than difficult. BS 5628: Part 3: 1985, Table 1 and Fig. 11, pages 36 and 37, compare local spell indices and the exposure categories given in CP 121: Part 1: 1973, the latter related to Lacy's annual mean DRI.

4.03 Water-excluding characteristics
Despite minor variations, materials can be classified into two groups:

- Impermeable (or impervious), i.e. those that do not allow the passage of water past their surfaces.
- Permeable, i.e. those that absorb water at their surfaces, some (or all) of which could eventually permeate through the material.

The use of *impermeable* materials at the exposed surface of an element (e.g. a wall cladding or roof finish) enables the principle of complete prevention to be applied, **4c**. Because of the greater 'run-off' of water on these surfaces, greater demands are placed on the design of joints in walls to ensure complete resistance to rain penetration. Only impermeable materials can be used on flat roofs and the slower 'run-off' of water on horizontal surfaces (including the possibility of ponding) makes special demands on the design and execution of the movement joints.

When *permeable* materials are used in walls, water may be excluded by the application of one of two principles.

1 Controlled penetration, i.e. allowing the penetration of water to a depth that will not cause problems, as in a solid wall. See **4a**.
2 Breaking capillary paths, as in the cavity wall, see **4b**.

Joints
For joints one of three methods may be used.
1 One line of defence, e.g. use of sealant at the exposed face.
2 Two lines of defence, e.g. providing for drainage of any water that may penetrate the seal at or near the exposed surface.
3 Joint geometry, e.g. profiling opposing faces of a joint to include anti-capillary and pressure-equalising grooves, and a baffle with provision for drainage.

4.04 Modes of water transmission
For transmission through an element as a whole:

1 The lateral penetration of water takes place through the combined effects of *capillarity* (through the pores of porous materials or along fine cracks at the interface of units of components) and *wind pressure*. The effects of the latter are increased as the pressure differences between two sides of an exposed layer (e.g. outer layers of a cavity wall) are increased.
2 The upward rise of water is entirely by capillarity.
3 The downward penetration is almost entirely by gravity and is increased by a head of water.

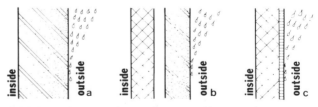

4 *Methods of water exclusion:* **4a**, *permeable wall;* **4b**, *cavity wall;* **4c**, *impervious wall. Shading indicates water penetration.*

For transmission through joints, in *Rainscreen Cladding*, page 10:

1 Kinetic energy
2 Surface tension
3 Gravity
4 Capillarity
5 Pressure assisted capillarity
6 Air pressure differentials

5 Further reading

- DD 93: 1984 *Draft for Development. Methods for assessing exposure to wind-driven rain.* Light on background but Prior (see below) makes up for that. Procedure explained in detail with worked examples. Includes new maps and related annual and spell rain roses.
- BRE Report, *Directional driving rain, indices for the United Kingdom – computation and mapping* (background to BSI Draft for Development DD 93) by M. J. Prior (BRE 1985). Explains the background missing in DD 93. Gives conversion of map values to DRI values. Essential reading for a full understanding of the new indices.
- Addleson, Lyall, *Materials for building,* Vol. 2. London, Illiffe Books, 1972. '3.02 Exposure' (pages 31–42) and '3.04 Exclusion' (pages 90–95).
  Detailed coverage of principles, extensively illustrated with diagrams and photographs. Section on driving-rain index overtaken by Lacy (below) and DD 93 (above).
- BRE Digest 127, *An index of exposure to driving rain* (March 1971). Succinct explanation of derivation of index. Better to use in conjunction with Lacy (below).
- Lacy, R. E., *Driving-rain index,* London, HMSO, 1976. Set of large-scale maps covering the whole of the UK with rules for adjusting map values for individual locations. Now overtaken by DD 93 (above).
- BRE *Protection from the rain* (1971, reprinted 1973).
  Well illustrated, selective and concise guide to the more common causes of rain penetration and includes descriptions of effective methods of protection.

6 Flat roofs

6.01 Causes of rain penetration

Rain penetration is usually caused by the effects of one or more of the following:

- splitting and tearing of the waterproof membrane generally, or near the internal angle of skirtings due to movements, foot traffic, and/or entrapped moisture
- cracking or detachment of the skirting and/or parapet, due to movements or loss of adhesion
- inappropriately located damp proof course trays in parapets.

6.02 Diagnostic checklist: membranes

- *Was the structure designed with movement joints?*
  Movement joints should be incorporated if any dimension of the roof exceeds 20 m in length, and/or at a change of direction in the roof.
  The principle of complete separation must be applied at the joint, and the detail adopted to ensure a waterproof joint must not provide restraint (see Study 24: Movements, diagram **12**). The junctions between the horizontal and vertical need special care and attention.

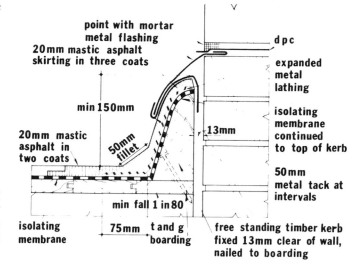

5 *Mastic asphalt laid on timber roof and mastic asphalt skirting on free-standing kerb (CP 114: Part 4: 1970).*

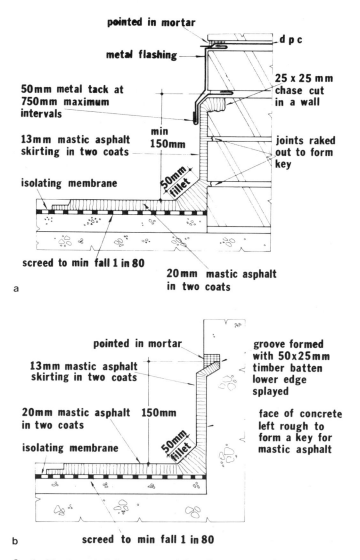

6a, b *Mastic asphalt laid on screeded roof and mastic asphalt skirting to brick wall, 6a; and to concrete wall, 6b.*

● *Was the surface protected with a solar reflective treatment?*
Solar protective treatment is always advisable but is
essential if thermal insulation is incorporated immediately below the waterproof membrane.

The efficacy of any treatment depends on surface
colour (white is best) during service (i.e. account must be
taken of the darkening effects of dirt – see Study 24:
Movement, 7.03, 7.04, page 139).

● *Was provision made to reduce the effects of differential movement
of the layers making up the roof construction?*
Drying out of porous materials before the membrane is
laid and/or provision for drying out after construction is
necessary to avoid problems associated with moisture
movement (see Study 21: 'Entrapped moisture', 5.03).
The location of the thermal insulating layer can influence
the thermal stability after construction – see Study 24:
Movements, 7.03, page 139.

● *Was the roof provided with adequate falls?*
All flat roofs including those with protected membranes
('inverted roof') should be laid to an achieved fall of at
least 1:80. The effects of structural deflection on the fall
actually achieved need to be considered – falls can easily
be nullified by roof sag.

6.03 Diagnostic checklist: asphalt skirtings

● *Was the skirting properly sealed at all internal angles?*
A two-coat 50 mm angle fillet is essential.
● *Was provision made for movement of timber?*
A free-standing timber kerb is essential, **5**.
● *Was the background adequately keyed?*
Brick joints need to be lightly raked out (i.e. 12 mm
depth) and brushed, before application of skirting.
Concrete, if not rough enough, can be either hacked or
treated with a proprietary bonding agent. All surface
laitence must be removed.

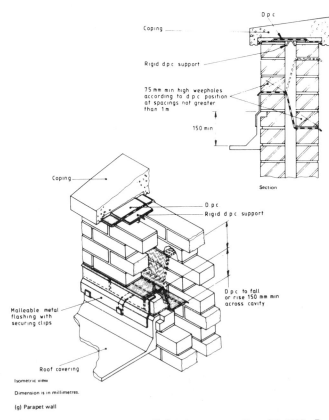

7 *Detail of coping to cavity wall forming parapet, from BS 5628: Part 3:
1985*

● *Were adequate chases provided for tucking in the top of the
skirting?*
In brickwork and concrete a splayed chase 25 × 25 mm
is required to allow tucking in and pointing, **6**.
● *Was the skirting protected with a solar reflective treatment?*
A solar reflective treatment is advisable with all warm
deck designs and essential with an inverted roof.

6.04 Diagnostic checklist: flashings
● *Do all flashings seal the top of skirtings effectively?*

6.05 Diagnostic checklist: dpcs/parapets
● *Were dpcs in parapets designed to ensure that rainwater is
prevented from penetrating through the joints in copings or
prevented from penetrating to the inside surface of walls?*
The joints between coping stones cannot be relied upon
not to open up a a result of movements. Down
penetration of water is by gravity so the location of the
dpc, including its relationship with the top of the skirting
to the roof, must take this into account, **7**.
● *Was allowance made for movement of the parapet?* (see Study
24: Movements).

6.06 Diagnostic checklist: projections through roofs
● *Were the skirtings adeqyate?*
150 mm is the minimum requirement – 200 mm with an
inverted roof.
● *Was the skirting properly sealed at all internal angles for large
projections such as rooflights?*
● *Was there sufficient space to lay the asphalt?*
Projections should be kept away from parapets and other
upstands sufficiently to allow the asphalt to be laid
properly
● *Were all cover flashings fixed?*
● *Do all flashings seal the top of the skirtings effectively?*
Cover flashings are essential, even with smaall projec-
tions such as pipes.
● *Was the skirting protected with a solar reflective treatment?*
A solar reflective treatment is advisable with all warm
decks and essential with an inverted roof.

6.07 Further reading: flat roofs
● CP 144: Part 3: 1970: *Roof coverings – built-up bitumen felt,
metric units* (with amendment AMD 2527, March 1978).
High-performance felts are not covered. Otherwise
comments as for Part 4 below.
● PSA, *Technical guide to flat roofing*, 1987. Comprehensive
with clear text and diagrams. Biased towards the
inverted roof. Supersedes PSA, *Inverted roofs, Technical
guide.*
● Tarmac's *Flat roofing, a guide to good practice.* Clearly
written with good diagrams. Covers principles as well as
practical application. The best reference of its kind.
● NHBC Practice Note 13, *Construction of flat roofs* (1981).
Very clear illustrations and tables incorporating the
latest knowledge.
● CP 144:Part 4: 1970: *Roof coverings – mastic asphalt, metric
units.* Contains the basic guidance necessary but some
parts require updating (e.g. recommendations for
ventilation of roof void are in conflict with BS 5250: 1975;
no reference to the protected membrane roof nor to use of
insulating materials such as expanded or extruded
plastics).

Table 11. Assessment of resistance to rain penetration
(A) Thickness of single-leaf walls with or without rendering

Exposure category	Minimum thickness of masonry (excluding rendering and finishes) (see note 1)				
	Clay and calcium silicate masonry		Concrete masonry		
	Rendered	Unrendered (see note 2)	Rendered (dense concrete)	Rendered (lightweight aggregate or autoclaved aerated concrete)	Unrendered (see note 2)
Very Severe	Not recommended. Cladding should be used				
Severe	mm 328	mm Not recommended	mm 250	mm 215	mm Not recommended
Moderate/Severe	215	Not recommended	215	190	Not recommended
Sheltered/Moderate	190	440	190	140	440
Sheltered	90	328	90	90	328
Very Sheltered	90	190	90	90	190

NOTE 1. Thickness of masonry is based on work sizes of masonry units i.e. tolerances are not included.
NOTE 2. Thicknesses of unrendered walls are based on the use of tooled joints filled completely with cement:lime:sand mortar.
NOTE 3. This table is intended to give guidance on the selection of forms of construction from the point of view of resistance to rain penetration only but other factors such as durability should be considered.

Table 11. *(concluded)*
(B) Factors affecting rain penetration of cavity walls

Factor affecting rain penetration	Increasing probability of rain penetration in the direction of the arrow			
Applied external finish (see 21.3.2.1)	Cladding	Rendering		Other (e.g. masonry paint, water repellent)
Mortar composition (see 21.3.2.3)	Cement:lime:sand		Cement:sand plus plasticizer or masonry: cement:sand	
Mortar joint finish and profile (see 21.3.2.4)	Bucket handle, weathered, etc.	Flush	Recessed, tooled	Recessed, untooled
Air space (clear cavity) (see 21.3.2.6)	Over 50 mm	50 mm	25 mm	None (see table 11(A))
Insulation (see 21.3.2.8)	None	Partial filling with 50 mm air space	Filled with type A insulant (25 mm cavity)	Filled with type B insulant (50 mm cavity)

NOTE 4. It is essential to read this table in conjunction with 21.3.2 and 21.3.3. In particular, the table does not take account of quality of workmanship (see 21.3.2.2) or the effect of architectural features (see 21.3.2.7).

8 *Facsimile reproduction from BS 5628: Part 3: 1985, showing recommended construction for the exclusion of rain under various conditions of exposure.*

Table 1. Mixes suitable for rendering (see clause 22)

Mix type	Cement:lime:sand	Cement:ready-mixed lime:sand		Cement:sand (using plasticizer)	Masonry cement:sand
		Ready-mixed lime:sand	Cement:ready-mixed material		
I	1:¼:3	1:12	1:3	—	—
II	1:½:4 to 4½	1:8 to 9	1:4 to 4½	1:3 to 4	1:2½ to 3½
III	1:1:5 to 6	1:6	1:5 to 6	1:5 to 6	1:4 to 5
IV	1:2:8 to 9	1:4½	1:8 to 9	1:7 to 8	1:5½ to 6½

NOTE. In special circumstances, for example where soluble salts in the background are likely to cause problems, mixes based on sulphate-resisting Portland cement or high alumina cement may be employed. High alumina cement should not be mixed with lime, ground limestone, ground chalk, silica flour or other suitable inert filler should be employed instead.

Table 2. Recommended mixes for external renderings in relation to background materials, exposure conditions and finish required (see clause 22)
NOTE. The type of mix shown in bold type is to be preferred.

Background material (see clause 19)	Type of finish (see clause 24)	First and subsequent undercoats			Final coat		
		Severe	Moderate	Sheltered	Severe	Moderate	Sheltered
(1) Dense, strong, smooth	Wood float	II or III	II or III	II or III	III	III or IV	III or IV
	Scraped or textured	II or III	II or III	II or III	III	III or IV	III or IV
	Roughcast	I or II	I or II	I or II	II	II	II
	Dry dash	I or II			II	II	II
(2) Moderately strong, porous	Wood float	II or III	II or III	III or IV	III	III or IV	III or IV
	Scraped or textured	III	III or IV	III or IV	III	III or IV	III or IV
	Roughcast	II	II	II	as undercoats		
	Dry dash	II	II	II			
(3) Moderately weak, porous*	Wood float	III	III or IV	III or IV			
	Scraped or textured	III	III or IV	III or IV			
	Dry dash	III	III	III	as undercoats		
(4) No fines concrete†	Wood float	II or III	II, III or IV	II, III or IV	II or III	III or IV	III or IV
	Scraped or textured	II or III	II, III or IV	II, III or IV	III	III or IV	III or IV
	Roughcast	I or II	I or II	I or II	II	II	II
	Dry dash	I or II	I or II	I or II	II	II	II
(5) Woodwool slabs‡	Wood float	III or IV	III or IV	III or IV	IV	IV	IV
	Scraped or textured	III or IV	III or IV	III or IV	IV	IV	IV
(6) Metal lathing	Wood float	I, II or III	I, II or III	I, II or III	II or III	II or III	II or III
	Scraped or textured	I, II or III	I, II or III	I, II or III	III	III	III
	Roughcast	I or II	I or II	I or II	II	II	II
	Dry dash	I or II	I or II	I or II	II	II	II

* Finishes such as roughcast and dry dash require strong mixes and hence are not advisable on weak backgrounds.
† If proprietary lightweight aggregates are used, it may be desirable to use the mix weaker than the recommended type (see 19.1.4).
‡ See 20.4 regarding special recommendations for the first coat.

9 *facsimile reproduction from BS 5262: 1976. Tables showing mixes suitable for rendering and suitability for background materials.*

- Mastic Asphalt Council and Employers Federation, *Roofing handbook* (May 1980).
  Follows recommendations given in CP 144: Part 4: 1970. Has excellent details – large and clear.
- FRCAB, *Roofing handbook,* June 1988. Not as comprehensive as PSA or Tarmac.
- BRE Report *Thermal insulation: avoiding risks,* 1989. Primarily related to thermal insulation and the avoidance of condensation but useful for showing how conflicts with rainwater exclusion should be resolved.
- BRE Digest 144, *Asphalt and built-up felt roofings: durability* (August 1972).
  Good summary of problem areas and appraisal of defects.
- Handisyde, C. C., *Everyday details* No. 18: 'Flat roofs: falls'.
  Good practical guidance.
- Handisyde, C. C., *Everyday details* No. 21: 'Parapets in masonry construction'.
  Good practical guidance on location of dpcs and treatment of skirtings in asphalt and built-up bitumen felt.

7 Pitched roofs

7.01 Causes of rain penetration

For each category below, rain pentration is usually caused by the effects of one or more of the causes given.
For small units (e.g. tiles and slates):

- end and/or side laps insufficient for pitch/degree of exposure
- sarking felt inadequately lapped, sags too much between rafters or at eaves, does not lap at verges, torn or inadequately formed around perforations through it (e.g. pipes, chimneys)
- poorly formed flashings and/or poorly formed soakers (or the soakers may have been omitted)
- units do not lie flat or are not in the same plane – the latter important for most interlocking types of tiles.
  For large units (e.g. profiled metal):
  - end and/or side laps insufficient and/or inadequately sealed for pitch/degree of exposure and/or sealant poorly applied or not fully compressed by fastener
  - too few fasteners used or incorrect fasteners used or fasteners incorrectly fixed (e.g. not vertical or do not compress sealant in laps sufficiently or washers distorted by overtightening)
  - thermal expansion – of dark-coloured claddings with thermal insulation below them especially
  - rooflight profiles do not match properly and/or end and/or side laps with metal incorrectly formed or too small
  - flashings and soakers around projections through the roof (e.g. rooflights with upstands or pipes) incorrectly formed (at laps and mitres in particular)

7.02 Diagnostic checklist: small units

- *Were the end and side laps related to the pitch and exposure of the roof to wind-driven rain?*
  Special care is needed with roofs of low pitch.
- *Was the sarking intended to be lapped and/or prevented from sagging too much and/or properly supported at the eaves so that water could drain into the gutter and/or 'sealed' around perforations through it?*
- *Were the flashings complete and watertight and/or were soakers included?*

Flashings around chimneys need special attention.
● *Were the rafters and other supporting timbers intended to ensure that the units lay flat and/or remained in the same plane?*
Special care is needed at eaves. Thermal insulation below the sarking should not cause the sarking to bulge

thereby preventing water to drain down the sarking or causing unevenness in the units – clip fixings do not hold down units such as slates as would conventional nail fixings.

7.03 Diagnostic checklist: large units

● *Were the end and side laps include the need for sealing them related to the pitch of the roof and its exposure to wind-driven rain?*
Special care is needed with roofs of low pitch – pitches below 10° are classifed as 'flat'!
● *Were the fasteners – type and frequency – related to the exposure of the roof to wind-driven rain, the need to compress any sealant in laps or the type of thermal insulation used below the cladding?*
The fasteners in roofs with high parapets around them need special consideration relative to unusual wind effects (e.g. vortices) in areas exposed to high winds especially.
● *Was allowance made for thermal expansion?*
Dark-coloured claddings with thermal insulation below them need special attention – the range of movements can be high.
● *Was the profile of rooflights related properly to the profile of the cladding?*
If at all possible, rooflights of any kind are better avoided. If they cannot, then they should be located as near as possible to the ridge of the roof.
● *Were the flashings and soakers complete and watertight?*
Excessively long flashing behind projections through the roof should be avoided. All large projections through the roof should be located as near as possible to the ridge of the roof.

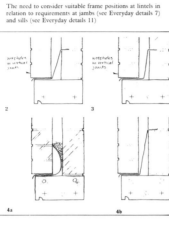

Although the general requirements for cavity wall lintels are well known, faulty detailing or workmanship still occurs, **1**. If faults at this position cause rain penetration the trouble is usually very difficult to cure.

In addition to ensuring protection against rain penetration general points to be kept in mind include:

Appearance (especially 'ears' at each side of opening).

Effect of lintels upon thermal insulation (ie 'cold bridges')

Provision of adequate fixings for frames, blinds and curtains

The need to consider suitable frame positions at lintels in relation to requirements at jambs (see Everyday details 7) and sills (see Everyday details 11)

Avoidance of rain penetration

2 is better than **1** because:

Top of flashing carried 50 mm into internal walls is secure.

Cement and sand fillet behind flashing reduces chance of damage during cavity cleaning.

Weep holes help to drain or dry the cavity.

but the acute angle at bottom of cavity makes cleaning difficult.

In theory, **3** has the advantages of **2** and simplifies cleaning without damage but the flashing is unlikely to finish in the position shown unless it is both flexible and stuck down.

With blockwork inner walls the normal (150 mm) rise of flashing will not reach the first blockwork joint, **4a**. Either increase the depth of flashing **4b** or start inner wall in brickwork or 150 mm course of blocks (usually available), **5**.

10a *Facsimile reproduction from the AJ Everyday details, 10, 'Cavity wall lintels',* **b** *and* **c** *Facsimile reproduction from BS 5628: part 3: 1985: Detailing for lintels.*

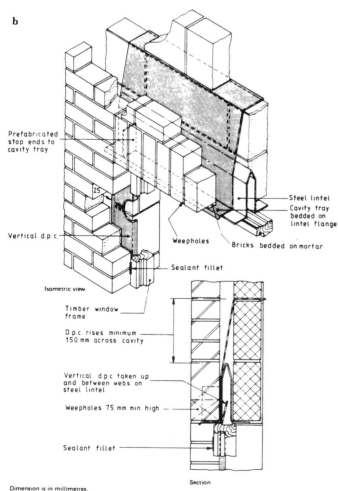

b

Prefabricated stop ends to cavity tray

Vertical d.p.c

Steel lintel

Cavity tray bedded on lintel flange

Weepholes

Bricks bedded on mortar

Sealant fillet

Isometric view

Timber window frame

D.p.c. rises minimum 150 mm across cavity

Vertical d.p.c. taken up and between webs on steel lintel

Weepholes 75 mm min. high

Sealant fillet

Section

Dimension is in millimetres.

(d) Steel lintel with separate d.p.c.

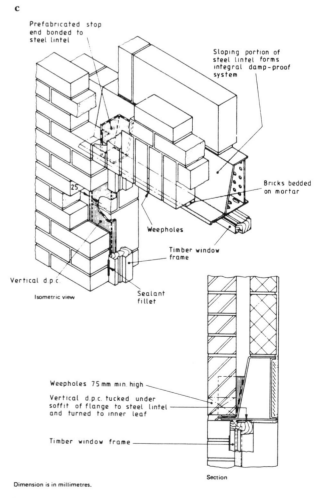

c

Prefabricated stop end bonded to steel lintel

Sloping portion of steel lintel forms integral damp-proof system

Bricks bedded on mortar

Weepholes

Timber window frame

Vertical d.p.c.

Sealant fillet

Isometric view

Weepholes 75 mm min. high

Vertical d.p.c. tucked under soffit of flange to steel lintel and turned to inner leaf

Timber window frame

Section

Dimension is in millimetres.

(e) Steel lintel with integral damp proof system

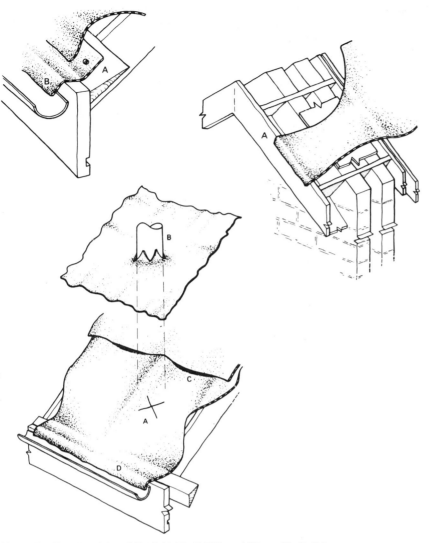

11 *DAS 9 Figs 2 and 3 and DAS 10 Fig 2 This and Figure 12, Building Research Establishment, Crown Copyright.*

Table 4. Recommended lap and angle of creep for slates : moderate exposure*

Rafter pitch not less than	Lap†	Length of slate			
		560 mm or longer	500 mm	460 mm	Less than 460 mm
		Angle of creep			
degrees	mm	degrees	degrees	degrees	degrees
45	65	26	26	26	26
40	65	26	26	26	26
35	75	26	26	26	29
30	75	32	32	32	32
25	90	36	40	47	—
20	115	48	56	65	—

*See BRE Digest 127, 'An index of exposure to driving rain'. For a more detailed assessment of local exposure, refer to Lacy R E, 'Driving-rain index', BRE Report, HMSO 1976.

†It is permissible to increase the lap in calculating the width of slate.

NOTE 1. These recommendations are minimum values which are more critical at rafter pitches below 30°, particularly where the thickness of the slate reduces the slope of the surface of the slate. At less critical pitches it would be normal to use a greater angle of creep, thus allowing for a loss of side lap where cutting of slates is necessary.

NOTE 2. Asbestos-cement slates. The recommendations for laps and side laps in tables 4 and 5 apply to asbestos-cement slates of 25° and below, except at roof pitches where the user should obtain confirmation of satisfactory performance at any lower pitches from the manufacturer.

Table 5. Recommended lap and angle of creep for slates : severe exposure*

Rafter pitch not less than	Lap†	Length of slate			
		560 mm or longer	500 mm	460 mm	Less than 460 mm
		Angle of creep			
degrees	mm	degrees	degrees	degrees	degrees
45	65	32	32	32	36
40	75	34	35	40	48
35	75	37	40	45	57
30	75	41	44	50	60
25	100	45	50	61	—
20	130	60	—	—	—

*See BRE Digest 127, 'An index of exposure to driving rain'. For a more detailed assessment of local exposure, refer to Lacy R E, 'Driving-rain index', BRE Report, HMSO 1976.

†It is permissible to increase the lap in calculating the width of slate.

NOTE 1. These recommendations are minimum values which are more critical at rafter pitches below 40°.

NOTE 2. Asbestos-cement slates. The recommendations for laps and side laps in double lap slating apply (see also table 4) except at roof pitches below 30°, where the user should obtain evidence of satisfactory performance from the manufacturer.

12 *BS 5534 Tables 4 and 5*

7.04 Further reading: small units

- BS 5534: Part 1: 1978: *Code of Practice for slating and tiling* (formerly CP 142: Part 2).
  Note: This code uses angles of creep to determine slate size and the side lap – see page 8.
- BRE DAS 9, *Pitched roofs: sarking felt underlay – drainage to roof* (Design).
- BRE DAS 9, *Pitched roofs: sarking felt underlay – drainage to roof* (Site).

7.05 Further reading: large units

- NFRC, *Profiled sheet metal roofing and cladding, a guide to good practice.* Better on application with good clear diagrams rather on performance.
- *AJ* series *Element Design Guide Roofs,* 4 Profiled sheet roofs by Peter Falconer, 1.04.87, pages 49–56. Good review with clear, helpful drawings of the critical parts.

8 Cavity walls

8.01 Causes of rain penetration

Rain penetration is usually caused by the effects of one or more of:

- the permeability of the units (e.g. some types of lightweight concrete blocks) and/or the mortar jointing
- the relative impermeabilility of the units and the type of jointing used with them (e.g. low-absorption bricks, notably perforated types, with recessed joints)
- cracking of rendered finishes
- bridging of the cavity (e.g. dirty wall ties and/or cavity fill)
- faulty dpcs and cavity trays.

8.02 Diagnostic checklist: walling units

- *Was the choice of the external walling unit and/or cladding and finish related to exposure to wind-driven rain?*
  Choice is dependent on such exposure – see **9**.
  Upper floors of multi-storey buildings can be severely exposed even in an area considered to be sheltered.
- Low-absorption bricks with recessed joints are vulnerable in severely exposed locations. Soldier courses are similarly vulnerable.

8.03 Diagnostic checklist: mortar mix, render mix, cavity fill

- *Was the mortar mix used related to exposure to wind-driven rain and the time of building?*
  See Study 25: Loss of adhesion, 4, page 147.
- *Was the mix for the render related to exposure to wind-driven rain and the characteristics of the background to which it was applied?*
  See Study 25: Loss of adhesion, 7, page 151.
- *Was the cavity filled?*
  The complete filling of cavities with an insulating fill of any kind has the effect of 'exaggerating' defects in the outer leaf or across the cavity that might otherwise not cause rain penetration. New brickwork tends to be leaky initially. In time it is better that the joints 'heal', thereby increasing the outer leaf's resistance to rain penetration. Agrément Certificates for some fills require that the fill is not installed for at least a year after the walls have been built. All Agrément Certificates for cavity fills give the conditions of exposure and/or height of walls for which given fills are suitable.

8.04 Diagnostic checklist: dpcs and cavity trays

- *Were these designed or built to extend beyond or fold down the outer face of the wall?*
  See reproduction from *Everday details* **11**.
- *Were slopes within the cavity properly supported?*
  Unsupported material may be damaged during construction and proper laps are difficult to achieve, **10**.

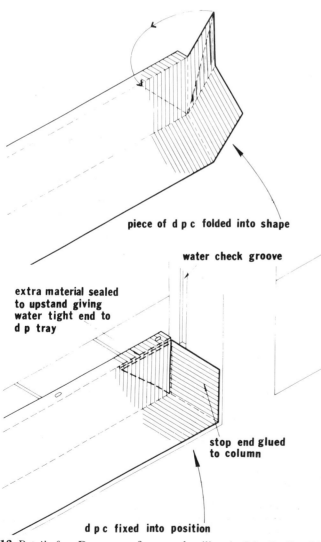

piece of d p c folded into shape

water check groove

extra material sealed
to upstand giving
water tight end to
d p tray

stop end glued
to column

d p c fixed into position

13 *Details from* Damp-proof course detailing *by John Duell and Fred Lawson.*

- *Were there end stops at the end of lintels?*
  With preformed steel lintels special attention should be paid to end stops, whether or not there is cavity fill, **10b** and **c**.
- *Were weepholes incorporated to drain the cavity?*
  Apart from allowing drainage of water flowing down the back of the outer leaf through weepholes, the presence of these helps to reduce the pressure difference across the two surfaces of the outer leaf and thus reduce in the amount of water penetration.
- *Were all junctions at returns (e.g. brick cladding/column junction) properly sealed?*
  It must be assumed that water will penetrate the outer leaf and such water must be excluded at returns – i.e. the problem must be considered in three dimensions, **11**.

8.05 Further reading

- BS 5628: Part 3: 1985: *British Standard Code of Practice for use of masonry,* 'Materials and components, design and workmanship' (formerly CP 121: Part 1: 1973). Fig. 12 shows details with preformed steel lintels and use of end stops, **10a, b, c**.
- BRE DAS 12, *Cavity trays in external cavity walls: preventing water penetration* (Design), December 1982. Figs 2 and 3 deal with lintels and mastering of jamb – nothing on end stops.
- BRE Report, *Thermal insulation: avoiding risks,* 1989, 2 Walls, pages 12–19. Useful summary for resolution of conflicts between the requirements of rainwater exclusion and thermal insulation. Includes a coloured map showing the six categories of exposure to wind driven rain according to BS 5628: Part 3 (page 13, 28).
- *Cavity insulated walls: Specifiers guide.* Good clear diagrams. Illustrates use of end stops.
- BS 5262: 1976: *Code of Practice for external rendered finishes* (formerly CP 211)
  One of the new style codes of practice that gives comprehensive guidance and includes reference to BRE Driving-rain index. There was an amendment in September 1976 (AMD 2103).
- Handisyde, C. C., *Everyday details,* Nos 10 and 15
  Good down-to-earth practical guidance on dpcs and cavity trays, including the junction with concrete upper floors.
- Duell and Lawson. *Damp-proof course detailing.* See particularly section 4, 'Detailing of DPC's and related works', pages 15*ff.,* and section 6 'Installation of dpcs' pages 25*ff.*
  An in-depth study of the selection, detailing and installation of dpcs, well written, comprehensive and extensively illustrated, mostly with very helpful three-dimensional drawings. Extremely useful (and timely) because of its emphasis on practical matters and proposals to put right details that have been found wanting. In many ways this supersedes most existing references.
- BRE Digest 236, *Cavity insulation.*
  See page 3: 'Rain penetration – experience and simulated tests', which suggests that blown-in rock fibre and polystyrene beads allow least water penetration.
  In very exposed situations the use of tile cladding or rendering will reduce the risk of penetration where other cavity insulation materials have been used.
- BRE Digest 277, *Built-in cavity wall insulation for housing.*
  Good for points that need special consideration in design and construction.

- BBA and others, *Cavity insulation of masonry walls – dampness risks and how to minimise them* (BBA, BRE, NFBTE & NHBC), November 1983. Comprehensive coverage of all the issues and includes details of the most common cavity fill materials in 'do and do not' data sheets.
- *Cavity insulated walls*, Practice Note 1, jointly by ACBA, BDA, C&CA, Eurosil-UK, NCIA and STA. Broadsheet format with does and do nots. Crisp and clear – good for checking on site. Specifiers guide below is more up to date and more informative. Concise and clear.
- *Cavity insulated walls*, Specifiers guide, jointly be ACBA, BPF, BDA, C&CA, Eurosil-UK, and NCIA. Addresses the main issues concisely with clear, crisp illustrations.
- *Agrément Certificates* for individual products (at least 75 certificates by mid 1989).
  Note: Most, if not all, Agrément Certificates give the limitations of the product or the circumstances of its use. It is important that these are read and interpreted for particular cases. See also Information Sheets below.
- BBA, Information Sheet 10, *Methods of assessing the exposure of buildings for cavity fill insulation* (revised edition, January 1983), relates to and formed the basis of BS 5618.
- BBA, Information Sheet 16, *Cavity wall insulation*, has important qualifications as to the suitability of certain walls for cavity fill, including the interpretation of local practice and for severe exposure conditions the increased risk of water penetration when low-porosity bricks and/or recessed joints are used. These qualifications were seldom included in certificates for particular products.
- BS 5618: 1985: *Code of Practice for the thermal insulation of cavity walls (with masonry or concrete inner and outer leaves) by filling with urea–formaldehyde foam systems.* A new edition of the code that updates suitable properties and tightens the installation procedure. Supersedes 1978 edition.
- BS 8208: Part 1: 1985: *Guide to the assessment of suitability of external cavity walls for filling with thermal insulants*, 'Existing traditional cavity construction'. Guidance on cavity walls not exceeding 12 m high or cavity walls without vertical members of the structural frame bridging the cavity. No guidance on design and construction of new work.
- BS 6676: Part 2: 1986: *Code of practice for installation of batts (slabs) filling the cavity construction*, gives criteria for design and construction of walls to be insulated and draws attention to precautions. Appendix give additional recommendations where walls exceed 12 m in height.
- BRE DAS 17, *External masonry walls: insulated mineral fibre cavity width batts – resisting rain penetration* (Site) (February 1983). Good on precautions.
- BRE DAS 79 *External masonry walls: partial cavity fill insulation – resisting rain penetration* (Site) (June 1986). On keeping cavities clean.
- BRE Digest 277, *Built-in cavity wall insulation for housing.* Good for points that need special consideration in design and construction.
- DOE, *Construction* 21, page 26, March 1977, 'Rain penetration of brickwork'.
  A simulation experiment which showed that walls built with mortars containing lime were the most effective in keeping out the rain.
- BRE Report, *Rain penetration through masonry walls: diagnosis and remedial measures*, 1988. Good summary of the essential features of cavity walls that lead to rainwater penetration – 1 Introduction and 2 Diagnosis, pages 1–3.

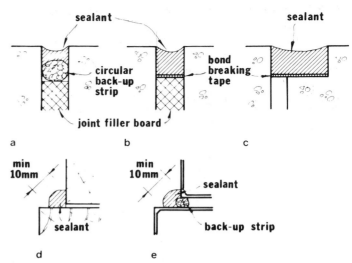

14 *Basic requirements of joints for sealants. Facsimile reproduction from* Manual of good sealant practice.

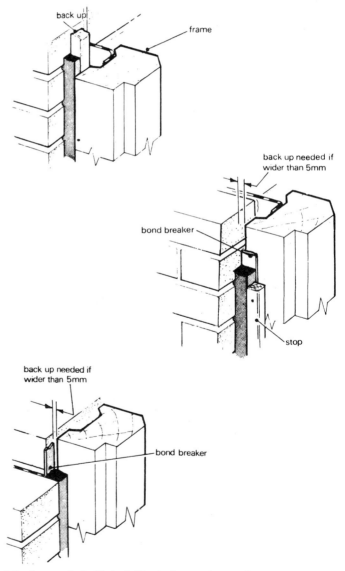

15a b c and d *(facing) Facsimile reproduction from BRE DAS 68, Figures 2 and 3 illustrating the requirements of a sealant at window/door jambs Building Research Establishment, Crown Copyright.*

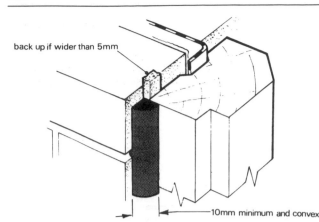

back up if wider than 5mm

10mm minimum and convex

15d

16a *Raked-out joints make good adhesion difficult to achieve.* **16b** *Ill-defined joints are difficult to seal properly.*

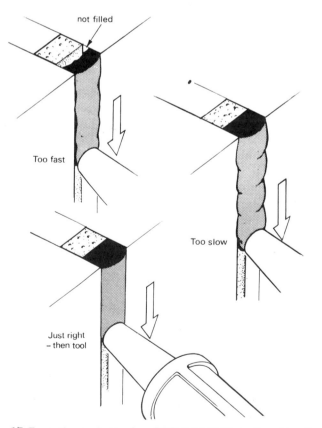

not filled

Too fast

Too slow

Just right – then tool

17 *Facsimile reproduction from BRE DAS 69 illustrating effect of speed of application on the finished sealant. Building Research Establishment, Crown Copyright.*

9 Around windows

9.01 Causes of rain penetration
Rain penetration is usually due to either or both of:
- inadequate seal at frame/reveal junction
- faulty dpcs and cavity trays.

9.02 Diagnostic checklists: seals
- *Was provision made for the effective application of sealing compounds?*
  Modern sealants are best applied to a well-defined joint (about 10 mm wide) with a back-up. The commonly used triangular fillet requires a width of 10 mm across the face and a convex surface. A back-up strip is essential if the gap between frame and reveal is wider than 5 mm, **15** and **17**.
  The movement of a standard domestic timber window frame can be as much as 4 mm. A rebate in the frame helps to reduce the strain on the sealant, **13b**.
  Good adhesion to rough or uneven surfaces (concrete or brickwork with raked out joints) is difficult to achieve, **16**.

9.03 Diagnostic cheklist: dpcs and cavity trays
- *Were the window frames located so that the vertical dpc could be effectively sealed against or within the frame to avoid bridging of moisture from the outer to the inner leaf?*
  As a guide, window frames should be positioned no nearer to the outside face than the line of the cavity.
- *Were the cavity trays designed and built to extend sideways to 200 mm beyond the opening to be protected?*
  See recommended detail **18**. When there is fill in the cavity and/or when preformed steel lintels are used, the projection of the tray beyond the opening does not provide complete protection. It is better to include end stops – see **10b** and **c**.
- *Were sufficient open vertical joints provided over the cavity trays?*
  At least one open joint per metre run required.
- *Were the vertical, lintel and sill dpcs designed and built to marry correctly?*
  The geometry of the dpcs can be complex, **8**. Correct marrying of all the dpcs requires three dimensional studies – in very difficult cases, models.

9.04 Further reading: windows
- BRE, *Protection from the rain*, April 1971 (reprinted 1973), pages 1 and 7.
  Main problems and possible solutions illustrated and concisely described.
- BS 5628: Part 3: 1985: *British Standard Code of Practice for use of masonry*, 'Materials and components, design and workmanship' (Formerly CP 121: Part 1: 1973). Fig. 12 shows details with preformed steel lintels and use of end stops.
- BRE DAS 68, *External walls: Joints with windows and doors – detailing for sealants* (Design) (December 1985). Clear, informative details in three dimensions.
- BRE DAS 69, *External walls: Joints with windows and doors – application of sealants* (Site) (December 1985). Clear, informative details in three dimensions. Fig. 3 helps to spot poor workmanship.
- BRE DAS 12, *Cavity trays in external cavity walls: preventing water penetration* (December 1982). Figs 2 and 3 deal with lintels and mastering of jamb – nothing on end stops.
- Herbert, M. R. W., *Window to wall joints*, BRE Current paper CP 86/74, September 1974, pages 1–3, 5–7.
  Problems analysed, faulty details and recommendation illustrated in detail. Some of the latter require redesigned

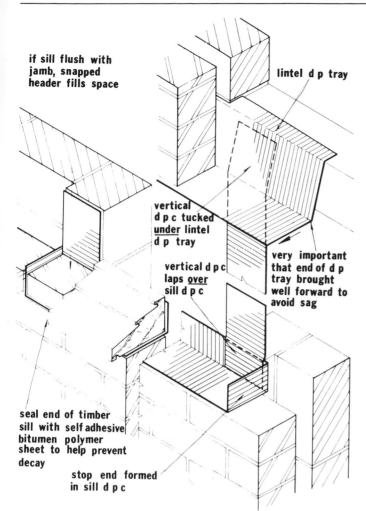

if sill flush with
jamb, snapped
header fills space

lintel d p tray

vertical
d p c tucked
<u>under</u> lintel
d p tray

vertical d p c
laps <u>over</u>
sill d p c

very important
that end of d p
tray brought
well forward to
avoid sag

seal end of timber
sill with self adhesive
bitumen polymer
sheet to help prevent
decay

stop end formed
in sill d p c

18 *(left) Correct detailing of and between dpcs is essential for maintaining a watertight assembly (from* Damp-proof course detailing, *by John Duell and Fred Lawson).*

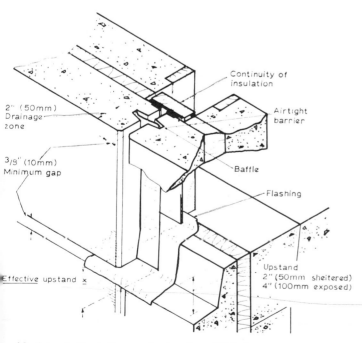

Continuity of
insulation

2" (50mm)
Drainage
zone

Airtight
barrier

3/8" (10mm)
Minimum gap

Baffle

Flashing

Effective upstand x

Upstand
2" (50mm sheltered)
4" (100mm exposed)

19 *(above) Facsimile reproduction from* BRS Digest 85 *(2nd series): drained joints between facade panels.*

steel window sections; others could place heavy demands on site assembly.

- Duell and Lawson, *Damp-proof course detailing.*
  See particularly section 4, 'Detailing of DPC's and related works', pages 15*ff*, and section 6, 'Installation of DPC's, pages 25*ff*.

10 Joints in concrete panels

10.01 Causes of rain penetration
In open-drained joints, **19**, rain penetration is usually caused by one or more of:

- incorrect location and/or lapping of vertical baffle
- inadequate projection of horizontal flashing
- inadequate air seal at back of panel, at junctions with columns especially.

In sealed joints (i.e. one line of defence) penetration is usually due to failure of the sealant (mainly loss of adhesion) caused by excessive movement and/or degradation.

10.02 Diagnostic checklist
For open-drained joints see above; for sealed joints see Study 25: Loss of adhesion, **8.03**, page 153.

10.03 Further reading
- Addleson, Lyall, *Materials for building*, Vol. 2. London, Illiffe Books, 1972, pages 134–155.
  Detailed coverage of joint design (including concrete panels). Fully illustrated with both diagrams and charts.
- BRE Digest 85, *Joints between concrete wall panels: open drained joints* (1971 edition).
  Main problems explained succinctly; source work for the open drained joint.
- Martin, Bruce. *Joints in building.*
  See particularly pages 115–117 for correct details
- *AJ* Series *Element Design Guide, External Walls,* 5 Precast cast concrete by Alan Brooks, 13.08.86, pages 35–40.
  Covers all aspects clearly. Well illustrated. 'Weather protection', pages 38 and 39, relevant to this study.

11 Thin metal or glass panels

11.01 Causes of failure
Rain penetration is caused primarily by failure of seal at frame due to movements (particularly thermal movement).

11.02 Diagnostic checklist
See Study 24: Movement, page 134.

11.03 Further reading
- Rostron, R. M., *Light cladding of buildings,* London, Architectural Press, 1964.
  Despite its age, still the best reference. Some parts, the details rather than the principles, have been overtaken by time and experience.
- Brookes, Alan, *Cladding of buildings,* London and New York, Construction Press, 1983. Detailed coverage of most systems.
- *AJ* Series *Element Design Guide, External Walls,* 3 Curtain walls by Barry Josey, 23.07.86, pages 47–65; 5 Metal panels by Alan Brookes, 6.08.86, pages 39–50. Both have good coverage with clear, helpful diagrams.
- Anderson J. M. and Gill, J. R., *Rainscreen cladding.* London, CIRIA/Butterworths, 1989. Useful for principles; parts on practice await further experience.

Study 23
Rising damp

1 Location of failures due to rising damp

Rising damp commonly occurs in:

- walls at or near ground level
- solid ground floor slabs, at junction with walls especially.

See Study 18: Wall/ground floor junctions.

2 Effects of rising damp

In walls, the effects are damage or deterioration of plaster and/or finishes; and increased surface condensation and mould growth.

In solid ground floor slabs, the effects are deterioration and/or loss of adhesion of floor finishes such as thermoplastic tiles.

3 Causes of rising damp

Main cause is bridging of the cavity, or perforation of dpc or dpm material.

4 Basic mechanisms

4.01 Ground water

Ground water reaching the base of a wall or the underside of a solid concrete ground floor slab will be soaked up by the porous materials of which these elements are composed by capillary action – like a wick. This action also accounts for the rise of ground water in cracks (i.e. very narrow spaces): this frequently occurs at the junction of wall and slab or through the slab itself. The height of the rise of moisture depends on the supply of water, on the pore structure of the materials, and on the rate of evaporation.

4.02 Hygroscopic salts

Ground water invariably contains dissolved salts which, having been drawn up a wall, tend to concentrate at the wall surface where the water evaporates. Some of these salts are hygroscopic and will absorb water from the air. The wall surface therefore tends to become damp whenever the air is humid (above about 70 per cent rh). Where existing walls are damp-proofed, the remedy may not be completely successful if provision is not made for:

- removal of salt-contaminated plaster
- prevention of the migration of salts within the masonry itself to the surface of new plaster. To meet the last requirement the undercoat should be basically a sand:cement render. Some proprietary plasters contain 'waterproofers'.

4.03 Prevention of rising damp

The rise of moisture is most effectively stopped by an impervious and continuous barrier using impervious materials such as asphalt, built-up felt and – more recently – plastic (polyethylene notably). Injection of water-repellent substances into walls (to line the pores of the materials) does not necessarily form a continuous barrier, although the barrier (which is a zone rather than a well-defined line as with the more usual dpc materials) formed is usually adequate to stop rising damp. The most

1 *Facsimile reproduction from BRS Digest 245 (2nd series) showing various ways in which the dpc may have been bridged, thus providing an alternative path along which water can rise up a wall.*

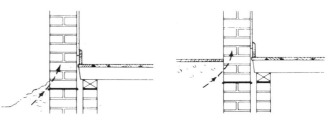

1a *Bridging by earth piled up against outside of wall.*
1b *Bridging by a path above dpc level (common with pavements).*

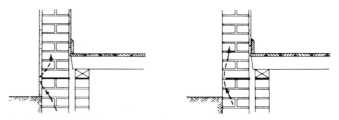

1c *Bridging of dpc by porous rendering.*
1d *Exposed edge of dpc bridged by pointing mortar.*

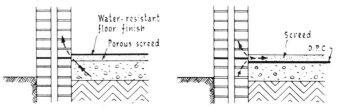

1e, f *Bridging by floor screed (which is usually porous, and may transmit moisture from wall below dpc level to that above).*

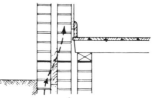

1g *bridging by mortar dropped between leaves of cavity wall.*

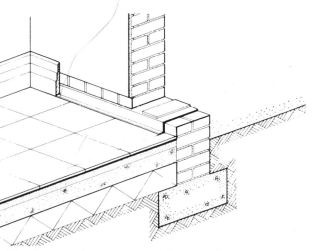

2a *New dpc inserted in existing wall. Dpc must be at least 150 mm above ground level. Damp-proof membrane should be turned down behind and under skirting or taken down to join floor. Note wedges inserted into sawn-out bed joint, over new dpc, to prevent wall settlement.*

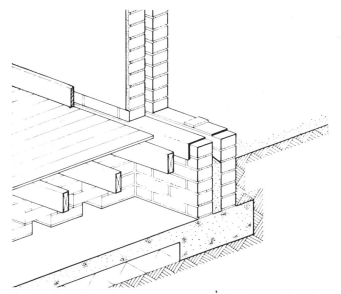

2b *In cavity walls the bed joints are sawn out and damp-proof membrane is inserted as in **2a**, but separately in each leaf. With timber floors, the membrane should be turned down below floor level. Additional treatment may be required to protect floor timbers.*

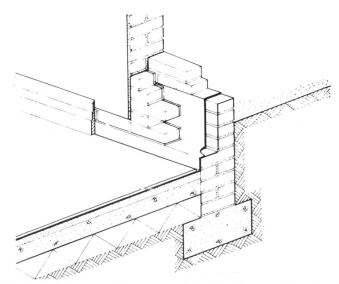

2c *Stepped dpc required where ground is above floor level. Insertion in existing wall necessitates access to interior of wall and extensive rebuilding. Alternatively, soil can be cut back.*

successful techniques appear to be those using materials mixable with water at the time of injection but which change, after injection, to form a water-repellent bond within the wall. Examples are a siliconate solution in water, or a silicone/latex mixture.

5 Diagnostic checklist

5.01 New work

- *Was provision made to avoid bridging the wall dpc?*
  The dpc may be bridged by mortar pointing or by rendering. The dpc must extend at least to the face of the wall or to the rendering, suitably broken at the line of the dpc, **1**.
- *Was the dpc or dpm perforated during laying?*
  The thinner dpc or dpm materials, such as polyethylene, are liable to perforation if the bedding or base is not smooth.
- *Was provision made for adequate linking of the dpc in the wall and the dpm in the ground floor slab?*
  See diagram **2a**, **2c**.
- *Was account taken to avoid cold bridging when either the wall or floor or both were insulated?*
  See diagram **3**.
- *Was account taken to avoid cold bridging when either the wall or floor or both are insulated?*
  See diagram **4**.

Table 2. Properties of flooring materials in relation to resistance to ground moisture penetration

Group	Material	Properties
A Finish and d.p.c. combined	Pitchmastic flooring Mastic asphalt flooring	Capable of resisting rising dampness without dimensional or material failure.
B May be used without extra damp protection	Concrete, Terrazzo Concrete or clay tiles	Capable of transmitting rising dampness without dimensional, material or adhesion failure.
	Cement/latex Cement/bitumen	Capable of partially transmitting rising dampness without dimensional or material failure and generally without adhesion failure.
	Wood composition blocks (laid in cement mortar). Wood blocks (dipped and laid in hot pitch or bitumen)	Capable of partially transmitting rising dampness without material failure and generally without dimensional or adhesion failure. *Only in exceptional conditions of site dampness is there risk of dimensional instability.*
C Not necessarily trouble-free without damp protection	Thermoplastic tiles (BS 2592) Vinyl asbestos tiles (BS 3260)	Capable of partially transmitting rising dampness through the joints without dimensional failure and generally without adhesion or material failure. Water penetration at the joints may result in decay at the edges in some conditions when ground water contains dissolved salts or alkalis.
D Reliable damp protection needed	Magnesite	Capable of transmitting rising dampness but adversely affected by water
	Flexible PVC flooring in sheet or tile form (BS 3261)	Impervious, but the flooring adhesive is sensitive to moisture.
	PVA emulsion cement	Impervious, but dimensionally sensitive to moisture (adhesive for tiles also sensitive to moisture).
	Rubber	Impervious, but prone to adhesion failure mainly through sensitivity of its adhesive.
	Linoleum	Sensitive to alkaline moisture attack through breakdown of bond and adhesive film.
	Cork Wood Chipboard	Acutely sensitive to moisture with dimensional or material failure.

3 *Facsimile reproduction of CP 102: 1973, table 2, relating type of floor finish to need for damp protection.*

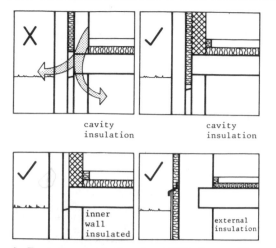

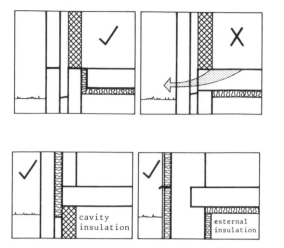

4 *Facsimile reproduction from BRE Report,* Thermal insulation: avoiding the risks, *1989, pages 22–27, illustrating the special needs when* *insulation is included in the wall or floor or both to reduce the risk of cold bridging. Building Research Establishment, Crown Copyright.*

5.02 Existing buildings

● *Was perished or salt-contaminated plaster removed and replaced (as it should have been) by a plaster that would prevent migration of hygroscopic salts to the surface?*
See above, 4.02.

● *Was the new plaster, to replace the cut-away perished plaster, brought down only to the line of the new dpc, and no lower? The dpc must not be bridged.*
This requirement is important whatever the dpc material, but special attention is needed to ensure that the band of a chemically injected dpc is not bridged by plaster. New skirtings may need to be deeper than the originals, to cover the unplastered wall surface below the new dpc; and should be fixed to treated grounds.

● *Was provision for adequate linking of the dpc in the wall and the dpm in the ground floor slab?*
Adequate linking could be difficult if the dpc in the wall was chemically injected.

6   Further reading

● CP 102: 1973: *Protection of buildings against water from the ground,* Section 3 'Damp-proofing of walls' and 'Damp-proofing of floors'.
The basic reference for new work, though it does not have many detailed drawings. Imminent replacement by a further part of BS 8102: 1990.

● Lawson, Fred, 'Dampness in buildings 1', *AJ* 10.3.71, pages 543–549 (CI/SfB 9 (12)).
Explains with the aid of excellent diagrams and photographs methods available for inserting a dpc in an existing wall. Causes of failures explained succinctly.

● BRE Digest 245, *Rising damp in walls: diagnosis and treatment,* January 1981.
Replaces Digest 27. The diagnostic test in Building Research Advisory Service Technical Information Leaflet 29 (*Diagnosis of rising damp*) is included. Chemical dpcs are more fully covered, incorporating information from BRAS TIL 36 (Chemical damp-proof courses for

walls), whereas the coverage of inserting physical dpcs is reduced.

● BRE Digest 77, *Damp-proof courses,* includes dpcs near the ground.

● BRE Digest 54, *Damp-proofing solid floors,* deals with effects of moisture on floor finishes and how to obtain protection from dampness rising from the ground.

● BRE DAS 22, *Ground floors: replacing suspended timber with solid concrete – dpcs and dpms.*

● BRE DAS 35, *Substructure: DPCs and DPMs – specification* (Design) (new edition, March 1985).

● BRE DAS 36, *Substructure: DPCs and DPMs – installation* (Site) (September 1983).

● BRE DAS 85, *Brickwalls: injected dpc's* (Design) (August 1986). Good summary of essentials.

● BRE DAS 86, *Brickwalls: replastering following dpc injection* (Design) (August 1986). Summary of essential points. An important aspect that is often overlooked or poorly executed.

● BS 6576: 1985: *Code of Practice for installation of chemical damp-proof courses,* includes recommendations for the chemical treatment of rising dampness in masonry walls and a description of diagnostic procedures and of installation.

● Duell, John and Lawson, Fred, *Damp-proof course detailing,* London, Architectural Press, 2nd edition 1983. This book deals with both new dpcs and the damp-proofing of existing buildings, and incorporates the *AJ* articles referred to under Fred Lawson's 'Dampness in buildings'.

● Richardson, B. A., *Remedial treatment of buildings*
See section 3.3, 'Rising dampness', page 97, for thorough discussion of diagnosis and treatment. He suggests that most rising dampness is due to a dpc being bridged, which is also the most easily remedied.

● BRE Report, *Thermal insulation: avoiding the risks,* 1989, page 22ff. With succinct text, illustrates clearly consequences of incorporating insulation in either floor or wall or both.

Study 24
Movement

1 Location of failures

Failure due to movements in elements and finishes occurs most frequently in the following locations.

1.01 Flat roofs
- Waterproofing membrane/screed/concrete slab, **1**
- Parapets, **2**
- Soffit to concrete slab (plaster finish).

1.02 Walls (external)
- Construction, **3, 4**
- Finishes/cladding – e.g. render, brick slips, **5**, tiles, mosaics, **6**
- Joints between cladding panels and (to a lesser extent) between window frames and reveal.

1.03 Walls (internal)
- Plaster.

1.04 Windows
- Joints in timber frames
- Paint finishes to timber frames.

1.05 Floors
- Finish, **7**.

2 Effects of movement

- Cracking of elements and finishes (see **1–4**)
- Detachment of claddings and finishes (see **5, 6**)
- Buckling of finishes, **7**, and components that in turn may lead:

At best
- to a change in appearance or

At worst
- to a breakdown of structural integrity
- to the ingress of water.

3 Types of movement

The types of movement *most* commonly responsible for cracking and detachment are due to:

- temperature changes (thermal movement) – occurs in all materials
- moisture content changes (moisture movement) – occurs in porous materials only.

Movements of types which are *less* commonly responsible for cracking and detachment may be due to:

1 *Example of major cracking in asphalt roof finish due to movement in substrate. Distinguished from minor cracking (due to other causes) in that crack is open and substrate usually visible between separated edges. (Photo GLC.)*

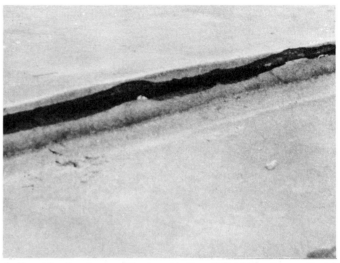

2 *Cracking of both concrete slab/beam, and brickwork above, by excessive movement along length of parapet.*

3 *Typical pattern of vertical cracking in brickwork due to movement.*

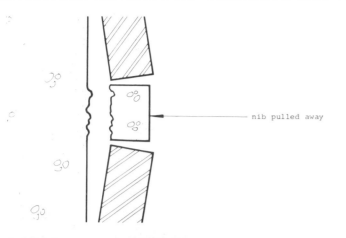

nib pulled away

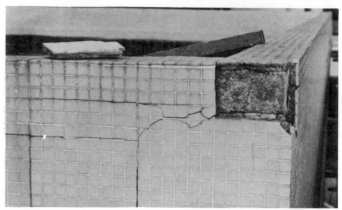

6 *Dislodgement of mosaic cladding. (Photo GLC.)*

4a, b *Dislodgement of both brick cladding and supporting concrete ledge. (Photo Building Research Establishment, Crown Copyright.)*

7 *Rippling of thin flooring due to moisture movement in screed below. (Photo Building Research Establishment, Crown Copyright.)*

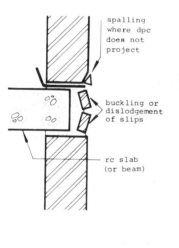

spalling where dpc does not project

buckling or dislodgement of slips

rc slab (or beam)

5a, b *Dislodgement of slips due to movement. (Photo BDA.)*

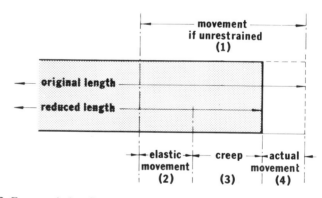

8 *Factors which influence stresses due to temperature or moisture changes (from* Principles of modern building, *Vol 1 p. 22).*

- deflection of structural members
- vibration (e.g. traffic, machinery and sonic booms*)
- chemical reactions (e.g. corrosion, sulphate attack, carbonation)
- other physical changes (e.g. ice or crystalline salt formation; loss of volatiles in mastics especially)
- movement in soils (e.g. settlement due to loading on silts and peaty soils particularly; moisture changes in clay soils; settlement due to mining subsidence).

* Note on wind effects: The vibrations or vortices caused by the wind in certain cases are not included in this study but have been noted in other studies. Failures of fixings or of conventional techniques of weather proofing in fully supported metal roofs (such as copper, aluminium and stainless steel) have been associated with high parapets around flat or low-pitched roofs or where the geometry of the roof is unusual. As yet there is no authoritative published guidance on problems that may arise. Importantly, wind-loading data for loading or other guidance are not entirely relevant. Individual cases have been the subject of special wind tunnel tests. Acknowledged wind experts should therefore be consulted where roofs have unusual features.

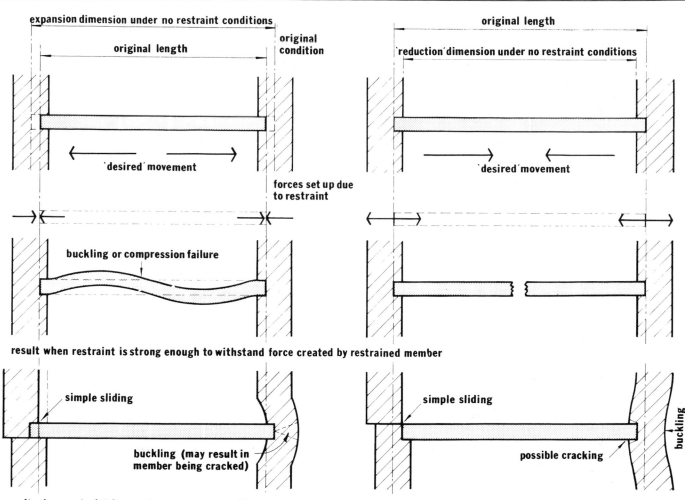

EXPANSION   **CONTRACTION**

9 *Diagrammatic representation of forces set up by restraint conditions; and two possible failures which may result.*

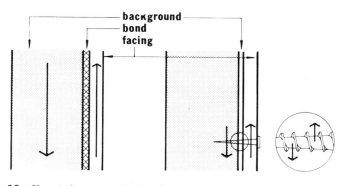

10a *Shear inducement in bond or fixing due to differential movement if bond or fixing does not remain intact.*

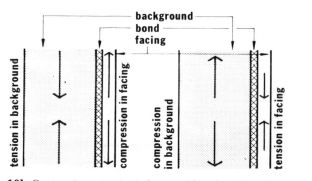

10b *Compression or tension inducement if bond or fixing does remain intact.*

4 Causes of failure

4.01 Dimensional changes

The changes (expansion or contraction) which result from changes in temperature and/or moisture content lead to failure if the combined effects of

- the magnitude of unrestrained movement of the material
- the modulus of elasticity of the material
- the capacity of the material to creep or flow under load
- the extent to which the material is restrained are

sufficient to induce stresses that:

- cause buckling
- exceed the strength of the material in compression, tension or shear
- exceed the strength of the bond between materials.

In practice the rate and amount of actual movement and the degree to which materials are restrained must be related to elastic movement and creep, **8**.

4.02 Basic mechanisms associated with movement

Provision of restraint against movement
Restraint may be provided by:

- the geometry of construction (e.g. corners; enclosing walls)
- fixings
- bonding or adhesion.

Tensile and compressive stresses

The circumstances likely to lead to tensile or compressive stresses in *restrained* members and the possible effects on the restraining members are shown in diagram **9**.

In *unrestrained* slabs, the combined effects of a thermal or moisture gradient across the slab and longitudinal movement (expansion or contraction) result in bowing of the slab as a whole (e.g. in roofs) or curling at the edges (e.g. in screeds).

Materials such as bricks (clay, concrete or calcium silicate) and cement-based products (plaster, render, screeds and unreinforced concrete) are weak in tension and are therefore prone to cracking when tensile forces are induced. Such forces commonly occur either within the complete thickness or at the surface of the material during shrinkage.

Shear stresses

The effect of restraint caused by adhesion or by fixings is usually inducement of shear stresses in the bond or in the fixing that results from differential movement. Alternatively, if the bond or fixing remains intact (wholly or partially), compression or tension will be induced in the adjoining surfaces, **10**.

5 Identification of cause

Identifying the precise cause or causes of failures due to movements is seldom simple. In this study, emphasis is given to movements resulting from changes in *temperature and/or moisture content*. In practice, failures often result from a combination of not only these movement but also of some of the other causes (listed earlier) operating at the same time or sequentially. For example:

- Shrinkage cracking in a *reinforced concrete member* allows the ingress of moisture to initiate corrosion of the mild steel reinforcement. The corrosion process then results in enlargement of the initial cracks or formation of new cracks (see Study 26: Corrosion 3, page 157).
- Shrinkage cracking of *external rendering* allows the ingress of moisture into the wall behind. The presence of moisture in the wall may result in loss of adhesion or further cracking of the rendering due to:
  (a) sulphate attack
  (b) carbonation
  (c) efflorescence
  (d) frost action.

6 Further reading

- Addleson, Lyall and Rice, Colin, *Performance of materials in buildings*, Butterworth-Heinemann, 1992. See '2.1 Cracking', pages 37–70. Broad-brush approach to assist in identifying the most likely problem areas and their solutions in principle. See also Addleson, Lyall, *Materials for building*, Vol 1, Iliffe Books, 1972 (OP), '2.03 Cracking in buildings' for explanations and illustrations of causes and effects of cracking on an analytical basis.
- BRE Digest 361, *Why do buildings crack*, September 1991. Deals with the causes of cracking rather differently from and replaces BRE Digest 75, *Cracking in buildings*, October 1966 that had a good summary of causes and effects of movements responsible for cracking.
- BRE Digest 217, *Wall cladding defects and their diagnosis*, 1978. Tabulated checklists for diagnosing defects, many of which are connected with inadequate allowance for movement.

- BRE Digests 227, 228, 229, *Estimation of thermal and moisture movements and stresses*, Parts 1, 2 and 3, 1979. Part 1 discusses movements, their sources, ways of designing to accommodate them, and the causes of deformation and stress. Part 2 is an analysis of thermal and moisture effects and includes tabulated data. Part 3 gives guidance on estimating deformations and associated forces and stresses.
- BS 5628: Part 3: 1985: *British Standard Code of Practice for use of masonry* (formerly CP 121: Part 1), 20 Movement in masonry, pages 31–35.
- BS 8200: 1985: *British Standard Code of Practice for design of non-loading external vertical enclosures of buildings*. Good for performance and principles – no details.

11 *Coefficients of linear thermal expansion of common building materials* $(\times 10^{-6}/°C)$.

Concrete	
Gravel aggregate	11.7
Lightweight aggregate	8.1
Limestone aggregate	6.0
Masonry blockwork	5.6 to 9.4
Clay products	
Bricks – clay: length	4.0 to 8.0
width	8.0 to 12
height	8.0 to 12
Bricks – calcium silicate: length	11 to 15
width	14 to 22
height	14 to 22
Stone	
Granite	8.5
Limestone	3 to 4
Marble	3.8
Sandstones	5 to 12
Plaster	
Gypsum	13.7
Perlite	9.3
Vermiculite	10.6
Glass	
Plate	9.1
Wood	
Across grain	50 to 60
With grain	3.8 to 6.5
Metals	
Aluminium and alloys	23.5
Brass	18.0
Bronze	19.8
Copper	16.9
Cast iron	10.6
Lead	28.6
Stainless steel	17.3
Mild steel	12.1
Plastics	
Acrylic – cast sheet	50 to 90
Polycarbonate	65
Polyester – 30% glass fibre	18 to 25
Rigid PVC	42 to 72
Phenolic	15 to 45
Expanded polyurethane	50 to 70
Foamed rigid polyruethane	20 to 70
Foamed phenolic	30 to 90
Expanded PVC	35 to 50

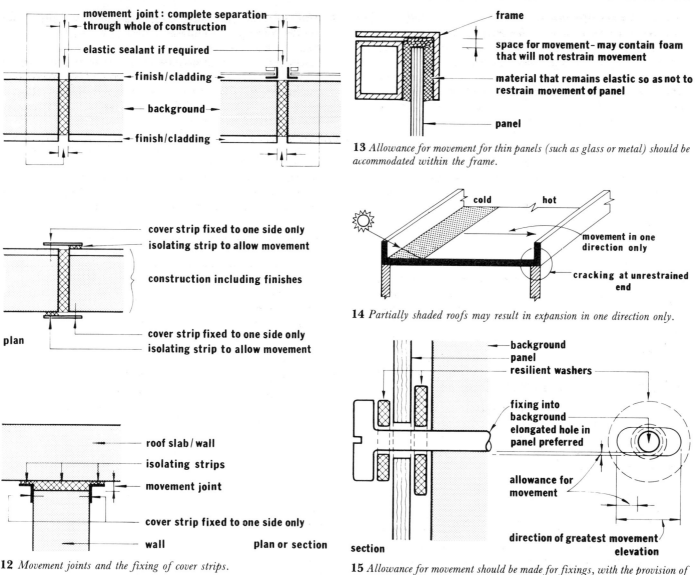

13 *Allowance for movement for thin panels (such as glass or metal) should be accommodated within the frame.*

14 *Partially shaded roofs may result in expansion in one direction only.*

12 *Movement joints and the fixing of cover strips.*

15 *Allowance for movement should be made for fixings, with the provision of elongated holes (where possible) in the direction of maximum movement.*

7 Diagnostic checklist (thermal)

7.01 Coefficients of linear thermal expansion

What are the coefficients of linear expansion of the materials involved? See **11**.

The coefficient of linear expansion is defined as the increase in unit dimension of a material for a temperature change of one degree and is expressed as:

$$\frac{\text{change in dimension*}}{\text{original dimension}} \times \text{temperature change}$$

- Coefficients apply to both expansion and contraction.
- Changes in dimension refer to unrestrained material.
- Most coefficients are relatively small† (e.g. 9.90×10^{-9} = 0.0000099) but this can be misleading as substantial changes in dimension can take place when the component dimension is large (i.e. lengths are long) or temperature change is large, or both. For example:

* Original and change in dimension must be expressed in the same units (i.e. both in metres or both in millimetres).
† Where a range of coefficients is given for a material, assessment in the case of failure should be based on the minimum and maximum value.

3 m length of aluminium/30°C change =
$24.30 \times 10^{-6} \times 3000 \times 30 = 2.2$ mm change in length.

3 m length of aluminium/65°C change =
$24.30 \times 10^{-6} \times 3000 \times 65 = 4.7$ mm change in length.

Or

30 m length of concrete/10°C change =
$9.90 \times 10^{-6} \times 3000 \times 10 = 3$ mm change in length.

30 m length of aluminium/30°C change =
$9.90 \times 10^{-6} \times 3000 \times 30 = 8.9$ mm change in length.

- Materials with high coefficients should always be suspect, particularly if they are thin and with low thermal capacity. The expansion of aluminium and some plastics (notably PVC) may be large for relatively short lengths (e.g. dimensional change for 2 m length of PVC for 30°C change could be as much as 4.3 mm) and the rate of movement may be rapid because of the combined effect of their high coefficient, thinness and low thermal capacity.

How has the actual movement been influenced or modified (apart from length and temperature rise) by:

- restraint
- heat transfer
- rate of heat conduction
- exposure
- movement joints?

These factors are described in detail in the sections that follow.

7.02 Restraint

If movement should have taken place (i.e. the design allowed for movement), was this movement restrained by incorrect detailing or by incorrect assembly of the construction as actually built?

- Movement joints should be formed in both background *and* in any finish or cladding applied to the background – i.e. separation to enable movement to take place must be continuous throughout the whole thickness of a construction. Any cover strips used to conceal a movement joint must be fixed to one side only and, if necessary, provision made for the other side to slide freely, **12**.
- Thin panels, e.g. glass and metal, held in frames require space for movement within the frame. Any fixing compound should remain elastic, **13**.
- The diameter of holes in panels, cover strips and trims should be larger than the fixings (e.g. screws and bolts) that pass through them and preferably elongated in the direction of maximum movement. Where possible, washers should be provided to enable the movement to take place, **15**.
- Areas (roof areas especially) that are shaded from sunlight for long periods may cause restraint. Expansion then takes place in one direction only in the exposed portion, and cracking is localised at the unrestrained end. Abutments and projections through the roof may also provide restraint that result in expansion of the roof taking place in one direction only, **14**.

Were bonded composite insulated panels balanced?

- Differences in the thermal expansion of thermal insulation (bonded with adhesives or by spraying – e.g. with polyurethane foams) and thin metal panels, particularly panels of a dark colour, induce restraints that result in bowing or distortion of the metal. The steep temperature gradient across the insulation is an important contributory factor. A balancing metal panel behind the insulation (i.e. so that the insulation is sandwiched between two metal panels) is required to resist the stresses causing the bowing or distortion of the panel.

7.03 Heat transfer

Which of the three heat transfer processes (i.e. conduction, convection and radiation) is dominant?

The temperature changes within materials, which then induce dimensional changes and therefore a tendency to movement, take place as a result of heat transferred by conduction, convection and radiation.

It is helpful, in investigating the pattern of temperature change underlying movement damage, to find out which of these transfers is dominant.

Radiation: external claddings and materials with *dark*-coloured surfaces tend to gain or lose heat more rapidly at their exposed surfaces by radiation (and therefore tend to

be subject to greater dimensional change) than materials with *light*-coloured surfaces.

Convection: heat loss by convection from the exposed surfaces of materials of any colour tends to be rapid if the surface of the material is *smooth* and if the air movement over the surface is both *rapid* and *cool*. However, the contribution of such convective heat loss to movement in building materials is likely to be insignificant.

Conduction: heat transfer between the surfaces of the cladding (or other external material) and the interior will occur only by conduction, and may be an important factor in movement. Such heat transfer between surfaces and interior will be strongly influenced by the presence of insulation layers, cavities and the like behind the external face of the material and by the precise position of such insulation.

If thermal insulation, for instance, is placed immediately behind a thin cladding or finish (e.g. behind glass, metal, built-up felt asphalt), then the cladding itself is likely to undergo large temperature fluctuations, and probably therefore also large dimensional changes, while the background material remains relatively stable. See the comparative graphs in **16**.

If the insulation is placed internally (e.g. the inside face of a wall or roof slab), then the *whole* of the construction (i.e. all the layers of which it is composed) will tend to undergo large temperature fluctuations.

The most stable thermal conditions, conducive to least movement, for the construction as a whole should be achieved when the insulation is placed *externally* over the finish or cladding, as in the 'protected membrane (inverted) roof' – see **16**.

Conclusion: Based on the above notes, a diagnostic check should consist of an evaluation both of the magnitude of temperature changes in particular parts of the construction and differential change between one part and another, leading to differential movements.

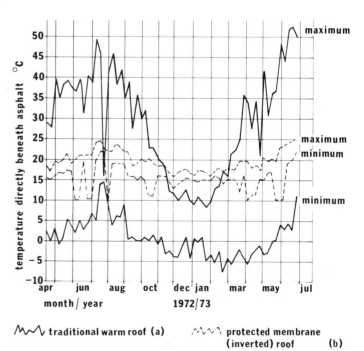

16 *Position of thermal insulation affects temperatures fluctuations in layers of construction. Above temperature variations were recorded immediately beneath asphalt in flat roofs — but in (a) insulation was beneath membrane; in (b) it was above. From GLC* Development and Materials Bulletin **90**, *December 1975.*

7.04 Exposure

Has the basic exposure (i.e. to ambient air temperatures) been modified by daily or seasonal fluctuations, by solar radiation or by the composition of the layers of material making up the construction?

- Exposure to solar radiation may increase the surface temperature of thin, low thermal capacity claddings (e.g. aluminium) seasonally by as much as 50°C and daily to between 15°C and 20°C, depending on surface colour and background. Rises in surface temperature due to solar radiation may be between 85°C for dark-coloured masonry and 140°C for dark metal, insulated behind.

Whitening of surfaces of all kinds can reduce the effect of solar radiation, and therefore surface temperatures, significantly:

- A 280 mm cavity brick wall exposed to bright sun (ambient air temperature 22.2°C) was 36.1°C unpainted, and 25°C when painted white and clean.
- A flat roof covered with asphalt on cork insulation was 45°C untreated, and 32.2°C when covered with white chippings and clean.
- In the absence of solar radiation, temperature patterns of wall surfaces of any orientation follow those of the outdoor air but, at night especially, that of a roof may be several degrees below the outdoor temperature when heat is lost by long-wave radiation to a clear, cloudless sky.

7.05 Further reading

- Addleson, Lyall, *Materials for building*, Vol. 4, London, Newnes-Butterworths, 1975. See section 4.03, 'Thermal movements'.
  Detailed study of all aspects, profusely illustrated with charts, diagrams and photographs.
- Building Research Current Papers, Design Series 36, Mack, G. W., *Demands on rubbers in buildings*, 1964.
  Useful information (not SI) on surface temperatures and movement at joints for a range of materials.
- BRS Current Paper 2/71, Ryder, J. F. and Baker, T. A., *The extent and rate of joint movements*, January 1971.
  Useful information on joint movements in metal claddings especially.
- BRE Digests 227, 228, 229, *Estimation of thermal and moisture movements and stresses*, Parts 1, 2 and 3.
  Part 1 discusses movements, their sources, ways of designing to accommodate them and the causes of deformation and stress. Part 2 is an analysis of thermal and moisture effects and includes tabulated data. Part 3 gives guidance on estimating deformations and associated forces and stresses.
- BS 5628: Part 3: 1985: *British Standard Code of Practice for use of masonry* (formerly CP 121: Part 1). 20 Movement in masonry, pages 31–35.
- BS 8200: 1985: *British Standard Code of Practice for design of non-loading external vertical enclosures of buildings.* Good for performance and principles – no details. 32 Thermal, moisture and structural movement, pages 23–26.

8 Diagnostic checklist (moisture)

8.01 Moisture movements of materials

What are the moisture movements of the materials involved?

TIMBER	MOISTURE MOVEMENT (% original length range 90% – 60% R.H.)		
	TANGENTIAL VALUE	(bar chart 3.0 2.0 1.0 — 1.0 2.0)	RADIAL VALUE
I SMALL MOVEMENT VALUES			
AFRICAN WALNUT	1.3		0.9
AFRORMOSIA	1.3		0.7
AGBA	1.3		0.6
BALSA	2.0		0.6
CEDAR, SOUTH AMERICAN	1.3		0.9
DOUGLAS FIR	1.5		1.2
HEMLOCK, WESTERN	1.9		0.9
IROKO	1.0		0.5
MAHOGANY, AFRICAN (Khaya ivorensis)	1.5		0.9
MAHOGANY, CENTRAL AMERICAN	1.3		1.0
OBECHE	1.25		0.8
OPEPE	2.0		1.1
PINE, YELLOW	1.7		0.9
ROSEWOOD, INDIAN	1.0		0.7
RHODESIAN, TEAK	1.6		1.0
SPRUCE, ENGLISH	1.3		0.9
TEAK	1.2		0.7
UTILE	1.8		1.6
WESTERN RED CEDAR	0.9-1.9		0.45-0.8
II MEDIUM MOVEMENT VALUES			
ASH	2.5		1.5
ELM, ENGLISH	2.4		1.5
JARRAH	2.6		1.8
KERUING	2.5		1.5
MAHOGANY, AFRICAN (Khaya grandifoliola)	1.8		1.3
OAK, ENGLISH	2.5		1.5
PARANA PINE	2.4		1.7
PINE, CARIBBEAN PITCH	2.6		1.4
PINE, SCOTS	2.1		0.9
POPLAR, BLACK ITALIAN	2.8		1.2
RED WOOD	2.2		1.0
SAPELE	1.1		1.3
SPRUCE, EUROPEAN	2.1		1.0
SYCAMORE	2.8		1.4
WALNUT, EUROPEAN	2.0		1.6
III LARGE MOVEMENT VALUES			
ASH, JAPANESE	3.5		1.5
BEECH	3.2		1.7
BIRCH, CANADIAN YELLOW	2.5		2.2
GURJUN	3.3		2.0
OAK, TURKEY	3.3		1.3
OLIVE, EAST AFRICAN	2.9		1.7
RAMIN	3.1		1.5
WATTLE, BLACK	3.5		1.2

17 *Comparative tangential and radial moisture movements in timber. Based on BRE Princes Risborough Laboratory, Technical Note 38,* The movement of timbers, *May 1969 (revised November 1982). Building Research Establishment, Crown Copyright.*

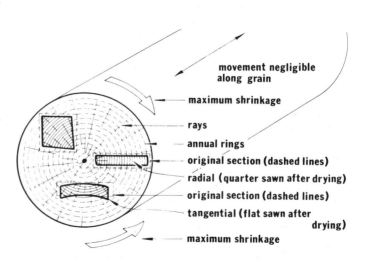

18 *Moisture movement is dependent upon direction of rings and therefore upon sawn section. Source:* Principles of modern building, *Vol. 1.*

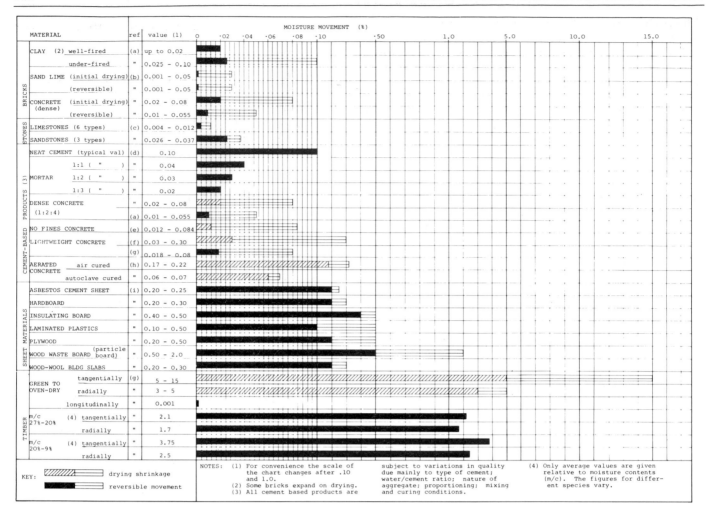

MATERIAL		ref	value (1)
BRICKS	CLAY (2) well-fired	(a)	up to 0.02
	under-fired	"	0.025 – 0.10
	SAND LIME (initial drying)	(b)	0.001 – 0.05
	(reversible)	"	0.001 – 0.05
	CONCRETE (dense) (initial drying)	"	0.02 – 0.08
	(reversible)	"	0.01 – 0.055
STONES	LIMESTONES (6 types)	(c)	0.004 – 0.012
	SANDSTONES (3 types)	"	0.026 – 0.037
CEMENT-BASED PRODUCTS (3)	NEAT CEMENT (typical val)	(d)	0.10
	MORTAR 1:1 (")	"	0.04
	1:2 (")	"	0.03
	1:3 (")	"	0.02
	DENSE CONCRETE (1:2:4)	"	0.02 – 0.08
		(a)	0.01 – 0.055
	NO FINES CONCRETE	(e)	0.012 – 0.084
	LIGHTWEIGHT CONCRETE	(f)	0.03 – 0.30
		(g)	0.018 – 0.08
	AERATED CONCRETE air cured	(h)	0.17 – 0.22
	autoclave cured	"	0.06 – 0.07
SHEET MATERIALS	ASBESTOS CEMENT SHEET	(i)	0.20 – 0.25
	HARDBOARD	"	0.20 – 0.30
	INSULATING BOARD	"	0.40 – 0.50
	LAMINATED PLASTICS	"	0.10 – 0.50
	PLYWOOD	"	0.20 – 0.50
	WOOD WASTE BOARD (particle board)	"	0.50 – 2.0
	WOOD-WOOL BLDG SLABS	"	0.20 – 0.30
TIMBER	GREEN TO OVEN-DRY tangentially	(g)	5 – 15
	radially	"	3 – 5
	longitudinally	"	0.001
	m/c 27%-20% (4) tangentially	"	2.1
	radially	"	1.7
	m/c 20%-9% (4) tangentially	"	3.75
	radially	"	2.5

KEY: ▨▨▨▨ drying shrinkage
      █████ reversible movement

NOTES: (1) For convenience the scale of the chart changes after .10 and 1.0.
(2) Some bricks expand on drying.
(3) All cement based products are subject to variations in quality due mainly to type of cement; water/cement ratio; nature of aggregate; proportioning; mixing and curing conditions.
(4) Only average values are given relative to moisture contents (m/c). The figures for different species vary.

19 *Chart from* Performance of materials in buildings, *p. 228, showing comparative moisture movements. References (a)–(g) are information sources, and are listed in the original.*

Moisture movement is defined as the change in unit dimension of a material expressed as a percentage of its original dimension, i.e.

$$\text{Moisture movement} = \frac{\text{change in dimension}}{\text{original dimension}} \times 100$$

Values of moisture movement may indicate either an expansion or a contraction of a material. They generally refer to unrestrained material, are usually based on the complete drying of small samples and therefore tend to suggest exaggerated movements. These should be used for diagnostic purposes, to assess the worst conditions that may have occurred, **19**.

In timber the amount of movement and mode of distortion is dependent on the direction of the annual rings; movement along the grain is negligible, **17, 18**.

What type of movement has taken place most?
Irreversible movement takes place in all materials requiring water for their use, i.e. all materials used in the wet trades. This occurs during the drying-out period mostly, causing shrinkage of the material if unrestrained.

The irreversible movement of timber during its initial drying out (i.e. before seasoning) is of little practical significance.

Reversible movement takes place in all materials and the amount and direction of movement is related to the moisture content of the material. Expansion takes place when water is given off by the material. Timber apart, materials exposed externally tend to have far greater reversible movements than those exposed internally.

How has the actual movement been influenced or modified by:
● restraint
● drying out
● prevailing conditions of humidity
● movement joints?

These are described in detail in the sections that follow.

8.02 Restraint

Timber
Was the timber sufficiently restrained by fixings or adhesives to reduce distortion or opening up of joints?

Plasters, renderings, mortars and screeds
Were these sufficiently restrained by their backgrounds?

All these materials require to be restrained during their drying-out shrinkage by sufficiently strong, non-shrinkable backgrounds through adhesion that is strong and uniform. Adequate restraint does not prevent cracking completely but does ensure that cracks are very fine and well distributed. Such cracking is acceptable. See Study 25: Loss of adhesion.

Good adhesion between successive coats (plaster and render) is essential. With screeds, maximum adhesion is obtained if the screeding is done soon after the concrete has been placed (say, within three hours of placing).

Restraint can be reduced by dividing the finish or screed into smaller bays. However, bays are not recommended with screeds, as curling that causes floor finishes to ripple can occur at the junction of the bays – it is usually more practical to make good the cracks.

Additional restraint (provided by metal lathing with an isolating membrane of building paper or polyethylene behind) is needed when plaster or renderings are continuous across backgrounds with different movements, **20**. In these circumstances it is often better to form a straight joint at the line of junction of the two backgrounds.

Masonry walls and partitions
Were these provided with too much restraint?

In contrast to finishes, walls and partitions are likely to crack vertically due to tensile stresses during shrinkage if they are restrained too much. Restraint may be provided by corners or by rigid connections (through bonding or metal ties and straps) to walls or frames.

Massive concrete elements
Was the reinforcement designed and distributed to control shrinkable cracking?

Composite cladding panels
Were these provided with a balancing sheet on the back of the panel?

The lesson of balance lies in plywood that has an unequal number of 'sheets'. The same principle should be applied to composite panels of other materials that are likely to undergo moisture movements. Bowing and distortion of the panels occurs if there is an imbalance.

8.03 Drying out

Time
Was sufficient time allowed for drying out before the application of finishes?

All processes requiring the use of water (e.g. brick or block laying, concreting, screeding, plastering and renderings) require a considerable time to dry out before finishes (e.g. plaster, renderings, floor finishes, paints) may be satisfactorily applied. There are three stages when porous materials dry out.

- Evaporation of free water from the surface: this takes place rapidly.
- Loss of water from the larger pores from within the material which takes longer because there is more water here and this water, as water vapour, has to find its way to the surface through a tortuous arrangement of pores. Drying out is concerned with this stage only.
- Loss of water from the fine pores or cells, which is exceptionally slow and could continue for several years. Because of this and factors that influence moisture equilibrium, materials are seldom absolutely dry, but dry enough for building purposes (e.g. timber is dry with moisture contents up to about 20 per cent plaster and lightweight concrete up to 0.2 per cent and 5 per cent, respectively). As a rough rule of thumb, materials may dry out at the rate of about 25 mm per month under good drying conditions (i.e. adequate but not excessive heat and ventilation) and without any moisture being added (from rain or high relative humidity of the air).

Checking moisture content
Was the moisture content checked?

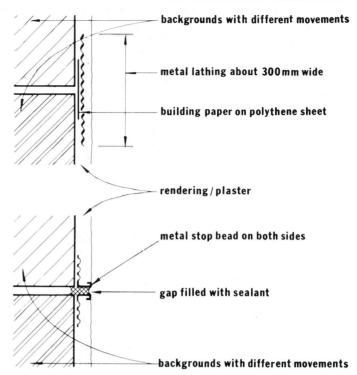

backgrounds with different movements

metal lathing about 300mm wide

building paper on polythene sheet

rendering / plaster

metal stop bead on both sides

gap filled with sealant

backgrounds with different movements

20 *Continuous plastering or rendering across backgrounds with different movements, requiring additional restraint and isolating membrane.*

It is impossible to determine dryness by merely inspecting or feeling the surface conditions. The interior of a material with a dry or apparently dry surface is invariably still damp. Moisture content may be checked with:

- electrical moisture meters (principally intended for testing timber but can give guidance on the condition of other materials if adequate precautions are taken), **21**.
- coloured indicator papers for surfaces to be painted
- hygrometers for use on walls and floors
- samples drilled out of walls and their moisture determined by weighing and drying or more conveniently on site with a Speedy moisture meter. Making good of drill holes is necessary (see 8.06, 'Further reading' for details of each method).

Speed of drying out
Was drying out accelerated in any way, either during construction or after completion?

Drying out cannot be accelerated successfully unless suitable precautions are taken, namely:

- use of heaters in moderation only
- avoidance of flueless heating burning gas or oil
- controlled use of dehumidifiers, especially when thin sections of timber are drying out
- provision of adequate ventilation (to take away moist air and bring in drier air) except when dehumidifiers are used. If the relative humidity is maintained above 70 per cent, mould growth will result; conditions approaching saturation will increase the moisture content of materials and thus exacerbate movement when drying out does take place.

Conditioning and protection
Were materials conditioned and were they adequately protected from absorbing moisture after incorporation?

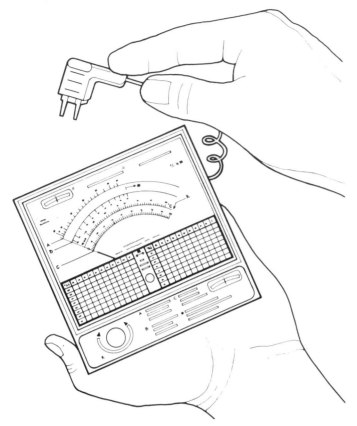

21 *Moisture meter; prongs are inserted into building fabric, and moisture content read off.*

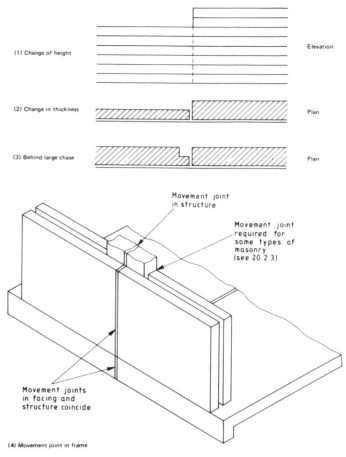

(1) Change of height　　　　　　　　　　　　　　　Elevation

(2) Change in thickness　　　　　　　　　　　　　　Plan

(3) Behind large chase　　　　　　　　　　　　　　Plan

Movement joint
in structure

Movement joint
required for
some types of
masonry
(see 20.2.3)

Movement joints
in facing and
structure coincide

(4) Movement joint in frame

(b) Locations where movement joints are likely to be required

23 *Facsimile of recommendations from BS 5628: Part 3: 1985.*

Table 2. Moisture content of timber for various positions

Position of timber in building	Average moisture content attained in use in a dried-out building (per cent of dry weight)	Moisture content which should not be exceeded at time of erection (per cent of dry weight)
Framing and sheathing of timber buildings (not prefabricated)	16	22
Timber for prefabricated buildings	16	17 for precision work, otherwise 22
Rafters and roof boarding, tiling, battens, etc.	15	22
Ground floor joists	18	22
Upper floor joists	15	22
Joinery and flooring (a) in buildings slightly or occasionally heated	14	14
(b) in continuously heated buildings	11-12	12
(c) in buildings with a high degree of central heating, e.g. hospitals	10	10
Wood flooring over heating elements	8-9	9

22 *Facsimile of table from BRE Digest 72 (Building Research Establishment, Crown Copyright).*

Bricks and blocks should have completed their drying shrinkage (with fired clay units their expansion) before incorporation into walls and timber should be at or near the moisture content level to be expected during service – something that is very difficult to achieve in practice.

All materials should be protected from absorbing ground moisture (by stacking off the ground) and rainwater (by covering with a waterproof covering) while being stored on site and during construction.

If materials have not been suitably conditioned or properly protected, extra time for drying out is essential especially before finishes are applied, **22**.

Composition of cement-based products
Were the cement-based products (mortars, plasters, renders and screeds especially) rich in cement and was an excessive amount of water used?

Shrinkage of all cement-based products increases as the cement and water content in the mix increases. (See also Study 25: Loss of adhesion.)

Shrinkable aggregates
Were shrinkable aggregates used in the concrete?

Cracking of concrete made with shrinkable aggregates has so far been confined to Scotland.

8.04 Prevailing conditions of humidity
What have been the prevailing conditions of humidity?
All materials adjust their moisture content according to the humidity of the air surrounding them until moisture equilibrium is attained.

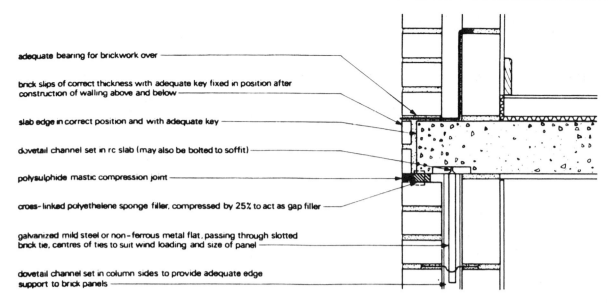

adequate bearing for brickwork over

brick slips of correct thickness with adequate key fixed in position after
construction of walling above and below

slab edge in correct position and with adequate key

dovetail channel set in rc slab (may also be bolted to soffit)

polysulphide mastic compression joint

cross-linked polyethelene sponge filler, compressed by 25% to act as gap filler

galvanized mild steel or non-ferrous metal flat, passing through slotted
brick tie, centres of ties to suit wind loading and size of panel

dovetail channel set in column sides to provide adequate edge
support to brick panels

24 *Facsimile from DOE,* Construction *14, June 1975.*

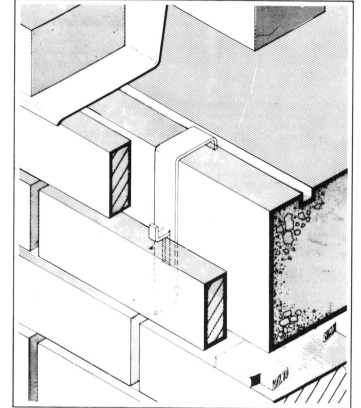

The absorption and giving off of moisture is not instantaneous. The time needed for moisture equilibrium to be reached depends on the nature of the material, its pore structure especially. Momentary changes in humidity do not have much effect on moisture movement.

Saturated materials continue to adjust their moisture contents until moisture equilibrium is reached, but this can take a long time and depends on the rate at which evaporation can take place at the surface (see 8.03, 'Drying out').

8.05 Movement joints
Were adequate movement joints provided, especially in elements composed of materials known to have large moisture movements?

- Walling. Movement joints, about 10 mm wide, should be provided at about every 6 m for walls of *concrete* units; every 7.5 m to 9 m for walls of *calcium silicate* units; about every 12 m for walls of fixed *clay* units.

These intervals are considerably less than those normally required for thermal movement.

25 *Facsimile reproduction of illustration from Brick Development Association Technical Note 9, showing support of brickwork by clip method.*

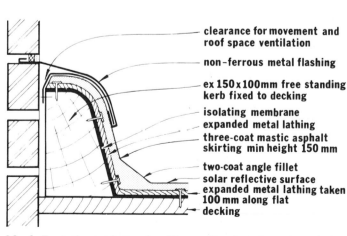

clearance for movement and roof space ventilation

non-ferrous metal flashing

ex 150 x 100mm free standing kerb fixed to decking

isolating membrane
expanded metal lathing

three-coat mastic asphalt skirting min height 150 mm

two-coat angle fillet

solar reflective surface
expanded metal lathing taken 100 mm along flat decking

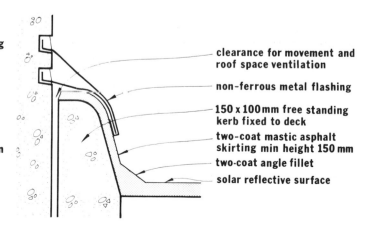

clearance for movement and roof space ventilation

non-ferrous metal flashing

150 x 100mm free standing kerb fixed to deck

two-coat mastic asphalt skirting min height 150 mm

two-coat angle fillet

solar reflective surface

26a, b *Facsimile reproduction from* The application of mastic asphalt, *showing preferred details for aplying mastic asphalt to free-standing kerbs.*

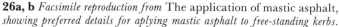

- External tiling and mosaics. Movement joints at least 6 mm wide should be provided 3 m apart vertically and horizontally.
- Brick cladding in concrete framed buildings. Special movement joint required at each storey height, **23–25**.
- Timber flat roofs – asphalt skirting. Special kerb detail required, **26**.

8.06 Further reading

- Addleson, Lyall and Rice, Colin, *Performance of materials in buildings*, Butterworth-Heinemann, 1992, '3.3 Moisture content', pages 223–277. Revision of Addleson, Lyall, *Materials for building*, Vol. 1 (pages 24–32) and Vol. 2 (pages 68–89), London, Iliffe Books, 1972.
- BS 5262: 1976: *External rendered finishes* (formerly CP 221).
  Comprehensive. Paragraphs 30 and 31 particularly relevant to resistance to cracking (pages 9 and 10).

Codes of practice
- BS 2568: Part 3: 1985: *British Standard Code of practice for use of masonry* (formerly CP 121: Part 1). Guidance on reducing cracking in 20 'Movement in masonry', pages 31–35 and selection of mortars in 23 'Selection of mortars', pages 63 and 64.
- BS 5385: Part 2: 1978: *Code of Practice for wall tiling: external ceramic wall tiling and mosaics.* (Superseded by CP 212: Part 2: 1966.)
  Provision of movement joints described succinctly (Clause 20, page 8.)

BRE Digests
- 35, *Shrinkage of natural aggregates in concrete,* with minor revisions, 1971.
  Concerned with problems encountered in Scotland.
- 79, *Clay tile flooring,* 1976.
  Details of precautions included.
- 104, *Floor screeds,* new edition 1973.
  Contains all the advice required.
- BRE Digest 362, *Building mortar,* October 1991. Gives recommendations for the composition and use of general purpose mortar and other specialised types of mortar. The recommendations reflect changes in British Standards and impending changes from British to European standards.
- 163, *Drying out buildings,* March 1974.
  Principles and methods for determining dryness (except drilling method – see Building Research Advisory Service, below) and modes of drying explained.

BRE other publications
- Building Research Advisory Service TIL 29, *Diagnosing of rising damp,* 1977. This technique is also covered in BRE Digest 245.
  Methods described are equally applicable to moisture content measurement in new buildings.
- BRE current Paper 7/75, *The independent core method – a new technique for the determination of moisture content.*
  Describes a technique for measuring accurately moisture content in porous walls. (Now also covered by BRE Digest 245.)
- DOE, *Construction,* **14/17**, Figure 7
  Gives correct detailing for movement joint in brick cladding to frame buildings.
- PSA, *Technical guide to flat roofing,* 1987, Figure 3.24, page 38 (repeated as Figure 3.129, page 125). Recommends movement joints in parapet copings 1.5 m from corners and thereafter at intervals of 4 m.

Study 25
Loss of adhesion

1 Finishes affected and locations

1.01 Most common failures
Loss of adhesion occurs most commonly in the following finishes and locations:

Finish affected	Location of finish	Section
Mortar	Walls	4
Plasters and renders	Walls and ceilings	5–7
Sealants and mastics	Joints, walls, and around windows	8
Tiles and mosaics	External walls	9
Paint	Timber windows	10
Built-up bitumen felt	Flat roofs	11
Asphalt skirtings	Flat roofs	12
Impervious sheet floor finishes	Solid ground floors	13
Flush movement joints	Asphalt flat roofs	14

2 Basic causes of loss of adhesion

Loss of adhesion may be due to either of the two main causes given below, but is usually due to a combination of both.

2.01 Inadequate preparation
Inadequate preparation of the background before application of the adhesive material. All adhesive materials require the background on which they are to be applied to be sound (i.e. not friable), clean and free from oil or grease. Some adhesive materials also require the background to be dry and/or with little suction.

2.02 Differential movement
Differential movement between the background and the adhesive materials.

2.03 Other causes
In addition to these two main causes, other reasons for loss of adhesion include:

- degradation of the adhesive material itself
- frost action
- efflorescence occurring in the background material.

3 Basic mechanisms of adhesion

3.01 Principles
The principles upon which good adhesion relies are fundamentally the same whatever materials are used. In all cases the adhesive material* is first applied in a liquid or

---

* The term 'adhesive material' is used for all agents of adhesion and not exclusively for materials commonly known or used as adhesives.

plastic state to a solid material (the background). After application the adhesive material changes state chemically and (mostly) physically. Some such as mortars, plasters and renders become solid; others such as the plastic and elastic mastics become plastic or elastic.

In most practical cases adhesion between two substances takes place by a combination of mechanical and specific adhesion. Mechanical adhesion is simply the interlocking of two substances; specific adhesion relies on the forces of attraction between the surfaces of the substances. It is not usually possible to determine precisely how much of the adhesion is due to mechanical means or to surface attraction. Adhesive materials such as mortars, plasters and renders rely for their adhesion largely on mechanical means, **1a**. For this reason their backgrounds must be provided with an adequate key, either naturally (e.g. porous and specially keyed materials) or mechanically (e.g. raked joints, roughened surfaces, keyed materials such as expanded metal or specially formulated bonding agents). Paints, sealants such as mastics and adhesives (in the commonly accepted sense) rely largely on specific adhesion although some form of key may still be advisable.

3.02 Conditions for successful adhesion
- *The ability of the adhesive material to 'wet' the background.*
  Two surfaces are always relevant: the surface of the adhesive material and the surface of the background. The presence on the background of substances that might interfere with wetting (e.g. dust, oil or grease) produces places of weakness in the adhesive bond. Once dried or chemically cured, the adhesive material will then have only limited areas of adhesion and can therefore be more easily removed from its background by mechanical means, vibration and, most importantly, by

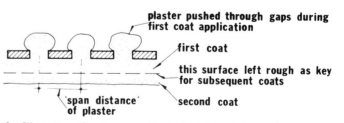

1a *Illustration of mechanical adhesion in lath and plaster.*

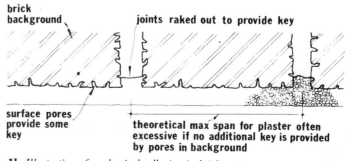

1b *Illustration of mechanical adhesion in brickwork.*

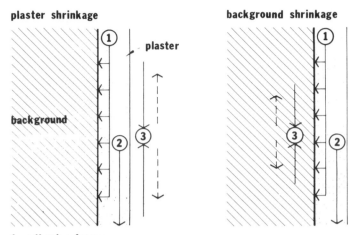

plaster shrinkage　　　　　background shrinkage

1 - adhesive forces
2 - forces caused by weight of material
3 - effects of shrinkage

2 *Effect of movement: When 1 is greater than 2 + 3 then adhesion is maintained. When 1 is less than 2 + 3 then adhesion is not maintained. Note: effect of expansion is the same in principle — but direction of expansion forces opposite to shrinkage forces shown, i.e. dotted lines. Diagrams based on* Materials for building, *Vol. 1, p. 68.*

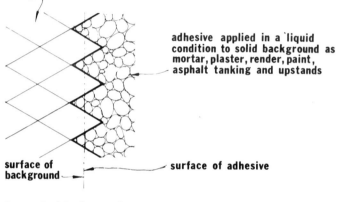

background

adhesive applied in a liquid condition to solid background as mortar, plaster, render, paint, asphalt tanking and upstands

surface of background　　　　　surface of adhesive

two materials, two surfaces

a

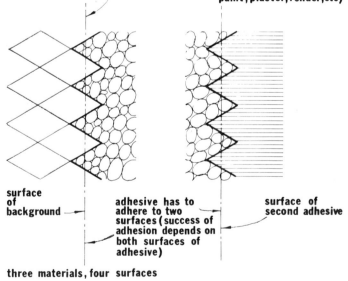

surface of adhesive(bonding coat, one coat paint, plaster, render, etc)

surface
of
background

adhesive has to adhere to two surfaces (success of adhesion depends on both surfaces of adhesive)

surface of
second adhesive

three materials, four surfaces

b

3 *Illustration of the importance of surfaces in adhesion, based on* Materials for building *Vol. 1, p. 69.* **3a**, *surface of background must adhere to surface of adhesive. In the case of three materials,* **3b**, *adhesive has to adhere to two surfaces at the same time. Adhesive surfaces indicated by jagged line.*

movement of the background or even the adhesive material itself, **2**.

● *Surface area available for adhesion*
One way of increasing the surface area is to roughen the background. The roughening also increases adhesion by mechanical means.

● *Maintenance of the liquid content of the adhesive material during and after application.*
Insufficient liquid within the adhesive material during application through suction by the backgroun may reduce the cohesive strength necessary in mortars, plasters and renders or interfere with the curing process necessary in paints and mastics. Backgrounds to receive mortars, plasters and renders are more commonly and conveniently wetted (care being required to avoid over-wetting), while those to receive paints and mastics may require to be sealed or primed (care being required to ensure that the sealer or primer is compatible with the adhesive material) — demonstrating the importance of considering the two surfaces in contact, **3**.

The most important requirement for self-adhesive materials (to which the mechanisms of specific adhesion are applicable) is the need for the background to which they are to be applied to have a smooth, even and plane surface, in addition to being clean and dry — requirements which are not always easy to achieve in practice.

3.03 Further reading

● Addleson, Lyall and Rice, Colin, *Performance of materials in buildings*, Butterworth-Heinemann, 1992, '2.2 Strength and the use of materials', '3 Adhesion', pages 108–116. Updated and extensively revised coverage of mechanisms and precautions in use previously in Addleson, Lyall, *Materials for building*, Vol. 1, London, Iliffe Books, 1972, pages 63–72.

● Marsh, Paul, *Fixings, fasteners and adhesives*, London and New York, Construction Press, 1984. Chapter 6 covers adhesives succinctly. Includes methods of application. Useful glossary of terms in Appendix 2.

● BRE Digests 211, 212, *Site use of adhesives*, Parts 1 and 2. Part 1 deals generally with basic adhesive types and the constraints of different backgrounds. Part 2 covers the selection of the right adhesive for the job in hand.

4 Mortars

4.01 Causes of failure

● Drying shrinkage of mortar.
● Differential movement (moisture or thermal but notably moisture) between the mortar and bricks or blocks.
Both result from the use of mortars that are cement rich and/or contain too much water during laying, **4**.

4.02 Effects of failure

● Cracking at the interface of the mortar and brick or block.
● Cracking of mortar joint.
Separately or together these result in reduced resistance of the wall to rain penetration, particularly in highly exposed situations. Loss of adhesion and/or cracking can be influenced by the width of the joints — the wider the joint, the greater the risk*.

* Complete avoidance of cracking is not possible in practice. The aim should be to ensure that the cracks are fine and well distributed. This is achieved by providing restraint to moisture movement through strong and uniform adhesion which was discussed in Study 24: 8.02, page 142.

Table 13. Durability of masonry in finished construction

(A) Work below or near external ground level

Masonry condition or situation	Quality of masonry units and appropriate mortar designations				Remarks
	Fired-clay units	Calcium silicate units	Concrete bricks	Concrete blocks	
A1 Low risk of saturation with or without freezing	Ordinary in (i), (ii) or (iii) or Special in (i), (ii) or (iii)	Classes 3 to 7 in (iii) or (iv) (see remarks)	$\geqslant$ 15 N/mm² in (iii)	(a) of block density $\geqslant$ 1500 kg/m³; (b) made with dense aggregate complying with BS 882 or BS 1047; (c) having a compressive strength $\geqslant$ 7 N/mm²; or (d) most types of autoclaved aerated block (see remarks) in (iii)	Some types of autoclaved aerated concrete block may not be suitable. The manufacturer should be consulted If sulphate ground conditions exist, the recommendations in **22.4** should be followed Where designation (iv) mortar is used it is essential to ensure that all masonry units, mortar and masonry under construction are protected fully from saturation and freezing (see clause **30** and clause **35**) The masonry most vulnerable in A2 and A3 is located between 150 mm above, and 150 mm below, finished ground level. In this area masonry will become wet and may remain wet for long periods of time, particularly in winter. Where Ordinary quality fired-clay units are used in A2 or A3, sulphate-resisting cement should be used (see **22.4**) Most Ordinary quality fired-clay units are not suitable for use in A3. The manufacturer should be consulted
A2 High risk of saturation *without* freezing	Special in (i) or (ii) *or* Ordinary in (i) or (ii) (see remarks)	Classes 3 to 7 in (ii) or (iii)	$\geqslant$ 15 N/mm² in (ii) or (iii)	As for A1 in (ii) or (iii)	
A3 High risk of saturation *with* freezing	Special in (i) or (ii) *or* certain Ordinary (see remarks) in (i) or (ii)	Classes 3 to 7 in (ii)	$\geqslant$ 20 N/mm² in (ii) or (iii)	As for A1 in (ii)	

(B) D.p.cs

	Fired-clay units	Calcium silicate units	Concrete bricks	Concrete blocks	Remarks
B1 In buildings	Ordinary or Special, which should have a water absorption $\leqslant$ 4.5 % measured as described in BS 3921, in (i)	Not suitable	Not suitable	Not suitable	Masonry d.p.cs can resist rising damp but will not resist water percolating downwards. If sulphate ground conditions exist, the recommendations in **22.4** should be followed. D.p.cs of fired-clay units are unlikely to be suitable for walls of other masonry units, as differential movement may occur (see **20.1**) Some Ordinary fired-clay units of water absorption $\leqslant$ 7 % may nevertheless be unsuitable for use in B2. The manufacturer should be consulted
B2 In external works	Ordinary (see remarks) or Special, which should have a water absorption $\leqslant$ 7 % measured as described in BS 3921, in (i)	Not suitable	Not suitable	Not suitable	

(C) Unrendered external walls (other than chimneys, cappings, copings, parapets, sills)

Masonry condition or situation	Quality of masonry units and appropriate mortar designations				Remarks
	Fired-clay units	Calcium silicate units	Concrete bricks	Concrete blocks	
C1 Low risk of saturation	Ordinary in (i), (ii) or (iii) or Special in (i), (ii) or (iii)	Classes 2 to 7 in (iii) or (iv) (see remarks)	$\geqslant$ 7 N/mm² in (iii)	Any in (iii) or (iv) (see remarks)	Walls should be protected by roof overhang and other projecting features to minimize the risk of saturation. However weathering details may not protect walls in conditions of very severe driving rain (see **21.3**). Certain architectural features, e.g. brickwork below large glazed areas with flush sills, increase the risk of saturation (see **22.5**). Where designation (iv) mortar is used it is essential to ensure that all masonry units, mortar and masonry under construction are protected fully from saturation and freezing (see clause **30** and clause **35**). Most Ordinary quality fired-clay units are not suitable for use in C2. The manufacturer should always be consulted. Where Ordinary quality fired-clay units are used in designation (ii) mortar for C2, sulphate-resisting cement should be used (see **22.4**).
C2 High risk of saturation	Certain Ordinary in (i) or (ii) (see remarks) or Special in (i) or (ii)	Classes 2 to 7 in (iii)	$\geqslant$ 15 N/mm² in (iii)	Any in (iii)	

(D) Rendered external walls (other than chimneys, cappings, copings, parapets, sills)

| Rendered external walls (other than chimneys, cappings, parapets, sills) | Ordinary in (i) or (ii) (see remarks) or Special in (i), (ii) or (iii) | Classes 2 to 7 in (iii) or (iv) (see remarks) | $\geqslant$ 7 N/mm² in (iii) | Any in (iii) or (iv) (see remarks) | Rendered walls are usually suitable for most wind-driven rain conditions (see **21.3**).
Where Ordinary quality fired-clay units are used, sulphate-resisting cement should be used in the mortar *and* in the base coat of the render (see **22.4**)

Where designation (iv) mortar is used it is essential to ensure that all masonry units, mortar and masonry under construction are protected fully from saturation and freezing (see clauses **30** and **35**) |

(E) Internal walls and inner leaves of cavity walls

| Internal walls and inner leaves of cavity walls | Internal or Ordinary or Special, in (i), (ii), (iii) or (iv) (see remarks) | Classes 2 to 7 in (iii) or (iv) (see remarks) | $\geqslant$ 7 N/mm² in (iv) (see remarks) | Any in (iii) or (iv) (see remarks) | Where designation (iv) mortar is used it is essential to ensure that all masonry units, mortar and masonry under construction are protected fully from saturation and freezing (see clauses **30** and **35**) |

(F) Unrendered parapets (other than cappings and copings)

Masonry condition or situation	Quality of masonry units and appropriate mortar designations				Remarks
	Fired-clay units	Calcium silicate units	Concrete bricks	Concrete blocks	
F1 Low risk of saturation, e.g. low parapets on some single-storey buildings	Ordinary in (i), (ii) or (iii) or Special in (i), (ii) or (iii)	Classes 3 to 7 in (iii)	$\geqslant$ 20 N/mm² in (iii)	(a) of block density $\geqslant$ 1500 kg/m³ (b) made with dense aggregate complying with BS 882 or BS 1047 (c) having a compressive strength $\geqslant$ 7 N/mm² (d) most types of autoclaved aerated block (see remarks) in (iii)	Most parapets are likely to be severely exposed irrespective of the climatic exposure of the building as a whole. Copings and d.p.cs should be provided wherever possible Some types of autoclaved aerated concrete block may not be suitable. The manufacturer should be consulted Most Ordinary quality fired-clay units are not suitable for use in F2. The manufacturer should be consulted. Where Ordinary quality fired-clay units are used, sulphate-resisting cement should be used (see **22.4**)
F2 High risk of saturation, e.g. where a capping only is provided for the masonry	Certain Ordinary in (i) or (ii) (see remarks) or Special in (i) or (ii)	Classes 3 to 7 in (iii)	$\geqslant$ 20 N/mm² in (iii)	As for F1 in (ii)	

(G) Rendered parapets (other than cappings and copings)

| Rendered parapets (other than cappings and copings) | Ordinary in (i) or (ii) (see remarks) or Special in (i), (ii) or (iii) | Classes 3 to 7 in (iii) | $\geqslant$ 7 N/mm² in (iii) | Any in (iii) | Single-leaf walls should be rendered only on one face. All parapets should be provided with a coping. Where Ordinary quality fired-clay units are used, sulphate-resisting cement should be used in the mortar *and* in the base coat of the render (see **22.4**) |

4 *Facsimile reproduction from BS 5628: Part 3: 1985 of mortar mixes.*

(H) Chimneys					
Masonry condition or situation	Quality of masonry units and appropriate mortar designations				Remarks
	Fired-clay units	Calcium silicate units	Concrete bricks	Concrete blocks	
H1 Unrendered with low-risk of saturation	Ordinary in (i), (ii) or (iii) or Special in (i), (ii) or (iii)	Classes 3 to 7 in (iii)	≥ 10 N/mm² in (iii)	Any in (iii)	Chimney stacks are normally the most exposed masonry on any building. Due to the possibility of sulphate attack from flue gases the use of sulphate-resisting cement in the mortar *and* in any render is strongly recommended (see **22.4**). Brickwork and tile cappings cannot be relied upon to keep out moisture indefinitely. The use of a coping is preferable
H2 Unrendered with high risk of saturation	Special in (i) or (ii) or Certain Ordinary (see remarks) in (i) or (ii)	Classes 3 to 7 in (iii)	≥ 15 N/mm² in (iii)	(a) of block density ≥ 1500 kg/m³ (b) made with dense aggregate complying with BS 882 or BS 1047 (c) having a compressive strength ≥ 7 N/mm² (d) most types of autoclaved aerated block (see remarks) in (ii)	Most Ordinary quality fired-clay units are not suitable for use in H2. The manufacturer should be consulted. Some types of autoclaved aerated concrete block may not be suitable for use in H2. The manufacturer should be consulted
H3 Rendered	Special in (i), (ii) or (iii) or Ordinary in (i) or (ii)	Classes 3 to 7 in (iii)	≥ 7 N/mm² in (iii)	Any in (iii)	
(I) Cappings, copings and sills					
Cappings, copings and sills	Special in (i) or certain Ordinary (see remarks) in (i)	Classes 4 to 7 in (ii)	≥ 30 N/mm² in (ii)	(a) of block density ≥ 1500 kg/m³ (b) made with dense aggregate complying with BS 882 or BS 1047 (c) having a compressive strength ≥ 7 N/mm² (d) most autoclaved aerated blocks (see remarks) in (ii)	Most Ordinary quality fired-clay units, including special bricks to BS 4729 and purpose made specials of this quality, are not suitable for use in I. Some autoclaved aerated concrete blocks also may be unsuitable. The manufacturer should be consulted. Where cappings or copings are used for chimney terminals, the use of sulphate-resisting cement is strongly recommended (see **22.4**). D.p.cs for cappings, copings and sills should be bedded in the same mortar as the masonry units

5 *(continuation of mortar mixes)*

Table 15. Mortar mixes					
		Mortar designation	Type of mortar (see note 2)		
			Cement:lime:sand (see note 3)	Air-entrained mixes (see note 5)	
				Masonry cement: sand (see note 3)	Cement:sand with plasticizer (see note 3)
			Proportions by volume (see note 4)	Proportions by volume	Proportions by volume
↑ Increasing strength (see note 1) and improving durability	Increasing ability to accommodate movements due to temperature and moisture changes ↓	(i) (ii) (iii) (iv) (v)	1:0 to ¼:3 1:½:4 to 4½ 1:1:5 to 6 1:2:8 to 9 1:3:10 to 12	1:2½ to 3½ 1:4 to 5 1:5½ to 6½ 1:6½ to 7	1:3 to 4 1:5 to 6 1:7 to 8 1:8
Direction of change in properties is shown by the arrows			Increasing resistance to frost attack during construction ⟶ Improvement in adhesion and consequent resistance to rain penetration ⟵		

NOTE 1. Where mortar of a given compressive strength is required by the designer, the mix proportions should be determined from tests following the recommendations of appendix A of BS 5628 : Part 1 : 1978.

NOTE 2. The different types of mortar that comprise any one designation are approximately equivalent in compressive strength and do not generally differ greatly in their other properties. Some general differences between types of mortar are indicated by the arrows at the bottom of the table, but these differences can be reduced (see **23.2.1**).

NOTE 3. The range of sand contents is to allow for the effects of the differences in grading upon the properties of the mortar. In general, the lower proportion of sand applies to grade G of BS 1200 whilst the higher proportion applies to grade S of BS 1200.

NOTE 4. The proportions are based on dry hydrated lime. The proportion of lime by volume may be increased by up to 50 % (*V/V*) in order to obtain workability.

NCTE 5. At the discretion of the designer, air entraining admixtures may be added to lime:sand mixes to improve their early frost resistance. (Ready mixed lime:sand mixes may contain such admixtures.)

6 *Facsimile reproduction from BS 5628: Part 3: 1985 showing minimum quality of fired-clay units and mortars for durability (key to mixes on 4).*

NOTE: There have been changes to the British Standard and changes from British to European standards are impending. See BRE Digest 362, *Building mortar*. This gives recommendations for the composition and use of general purpose mortar and other specialised types of mortar. The recommendations reflect changes in British Standards and impending changes from British to European standards.

Table 1. Mixes suitable for rendering (see clause 22)

Mix type	Cement:lime:sand	Cement:ready-mixed lime:sand		Cement:sand (using plasticizer)	Masonry cement:sand
		Ready-mixed lime:sand	Cement:ready-mixed material		
I	1:¼:3	1:12	1:3	—	—
II	1:½:4 to 4½	1:8 to 9	1:4 to 4½	1:3 to 4	1:2½ to 3½
III	1:1:5 to 6	1:6	1:5 to 6	1:5 to 6	1:4 to 5
IV	1:2:8 to 9	1:4½	1:8 to 9	1:7 to 8	1:5½ to 6½

NOTE. In special circumstances, for example where soluble salts in the background are likely to cause problems, mixes based on sulphate-resisting Portland cement or high alumina cement may be employed. High alumina cement should not be mixed with lime, ground limestone, ground chalk, silica flour or other suitable inert filler should be employed instead.

Table 2. Recommended mixes for external renderings in relation to background materials, exposure conditions and finish required (see clause 22)

NOTE. The type of mix shown in bold type is to be preferred.

Background material (see clause 19)	Type of finish (see clause 24)	First and subsequent undercoats			Final coat		
		Severe	Moderate	Sheltered	Severe	Moderate	Sheltered
(1) Dense, strong, smooth	Wood float	II or III	II or III	II or III	III	III or IV	III or IV
	Scraped or textured	II or III	II or III	II or III	III	III or IV	III or IV
	Roughcast	I or II	I or II	I or II	II	II	II
	Dry dash	I or II	I or II	I or II	II	II	II
(2) Moderately strong, porous	Wood float	II or III	III or IV	III or IV	III	III or IV	III or IV
	Scraped or textured	III	III or IV	III or IV	III	III or IV	III or IV
	Roughcast	II	II	II	as undercoats		
	Dry dash	II	II	II			
(3) Moderately weak, porous*	Wood float	III	III or IV	III or IV			
	Scraped or textured	III	III or IV	III or IV	as undercoats		
	Dry dash	III	III	III			
(4) No fines concrete†	Wood float	II or III	II, III or IV	II, III or IV	II or III	III or IV	III or IV
	Scraped or textured	II or III	II, III or IV	II, III or IV	III	III or IV	III or IV
	Roughcast	I or II	I or II	I or II	II	II	II
	Dry dash	I or II	I or II	I or II	II	II	II
(5) Woodwool slabs*†	Wood float	III or IV	III or IV	III or IV	IV	IV	IV
	Scraped or textured	III or IV	III or IV	III or IV	IV	IV	IV
(6) Metal lathing	Wood float	I, II or III	I, II or III	I, II or III	II or III	II or III	II or III
	Scraped or textured	I, II or III	I, II or III	I, II or III	III	III	III
	Roughcast	I or II	I or II	I or II	II	II	II
	Dry dash	I or II	I or II	I or II	II	II	II

*Finishes such as roughcast and dry dash require strong mixes and hence are not advisable on weak backgrounds.
†If proprietary lightweight aggregates are used, it may be desirable to use the mix weaker than the recommended type (see 19.1.4).
‡See 20.4 regarding special recommendations for the first coat.

7 *Facsimile reproduction from BS 5262: 1976 showing (table 1, top) mixes suitable for rendering; and (table 2, bottom) recommended mixes for external renderings in relation to background materials, exposure conditions and finish required.*

4.03 Diagnostic checklist
Was the mortar mix related to:

- *The type of construction?*
  Richer mixes are required for external walls and for brickwork; leaner mixes for internal walls and blockwork, **5**.
- *The conditions of exposure?*
  Richer mixes are required for retaining walls, parapets, free-standing walls or below dpc whatever the condition of exposure, **5**.
- *Time of construction?*
  Richer mixes are required during the winter building, particularly where there is an early frost hazard, **6**.

4.04 Further reading
- BRE Digest 362, *Building mortar* (Supersedes Digest 160). Gives recommendations for the composition and use of general purpose mortar and other specialised types of mortar. The recommendations reflect changes in British Standards and impending changes from British to European standards.
- BS 5628: Part 3: 1985: *British Standard Code of Practice for use of masonry*, 22 Selection of mortars, pages 63 and 64.

5 Gypsum plaster finishing coats on lightweight concrete blockwork

5.01 Causes of failure
Combination of:

- weakness in the bond between plaster and background or between successive layers of plaster, and
- differential movements of the plaster and background (usually drying shrinkage but could be thermal expansion) resulting in shear stresses sufficient to break a weak bond.

5.02 Effects of failure
- Shelling and bulging of the finishing coat which normally manifest themselves about 6 to 9 months after completion of plastering.

5.03 Diagnostic checklist
The items listed below are concerned with drying shrinkage. Movement from the greater drying shrinkage of the blockwork and undercoat plaster relative to the finishing is important. Shrinkage of lightweight concrete blockwork occurs as it dries out (the wetter the blocks, the greater the shrinkage); cement-based undercoats shrink with the blockwork but finishing coats of gypsum shrink very little on drying out.

- *Were the blocks dried to a moisture content below 30 per cent of that when fully saturated?*
- *Were the blocks protected from rain during transit and on site?*
- *Were the blocks stacked and covered on site to allow drying to continue?*
- *Was the walling protected during construction?*
  If not, and the blocks became wet, were they allowed to dry out before plastering or rendering?
  In the absence of protective measures to ensure dryness of the blocks before, during and after laying, the only course open is to allow the blocks to dry out before plastering or rendering.
- *Was a weak mortar with a mix appropriate to time of year used?* See figures **4** to **6**.
  Appropriate weak mortar mixes include:

  1:1:5–6 (cement:lime:sand) in winter
  1:1:8–9 (cement:lime:sand) in summer for external work protected during construction, or
  1:1:8–9 (cement:lime:sand) in winter
  1:1:10–12 (cement:lime:sand) in summer for internal work protected during construction.
  Alternatively, masonry cement:sand, or plasticised cement:sand mortars of equivalent mixes.
- *Was a finishing coat of gypsum plaster applied to a cement-based undercoat before the latter was dry?*
  The undercoat must be allowed to dry sufficiently to develop some suction. The common practice to apply finishing coats within a day of applying the undercoat is not advised.

5.04 Further reading
- DOE, *Construction* **2**/45, June 1972.
  Basic analysis of causes with advice on new work. List of references.
- Building Research Advisory Service TIL 14, *Shelling of plaster finishing coats.*
  The problem arises particularly where a cement rich undercoat is applied to a blockwork background. Whilst these cement undercoats shrink sympathetically on drying, gypsum finishing coats are more stable and therefore, if the bond with the undercoat is weak, will shell off under the shear stresses set up. To reduce the risk of failure the application of the finishing coat should be delayed until the undercoat has dried out. Under test, it was found that seven days' drying as opposed to 24 hours made a significant improvement.
- BS 5492: 1977: *Code of Practice for internal plastering.*
  Comprehensive, but see particularly clause 26, p. 10, on combinations of undercoats and finishing coats.

6 Plastering on dense concrete/soffits

6.01 Causes of failure
Combination of:

- inadequate control of suction of the concrete background, resulting in a weak bond, and
- differential movement, usually from thermal changes, causing shearing forces to be set up at the interface of the two materials (plaster and concrete) sufficient to break the weak bond.

6.02 Effects of failure
- Complete separation of the plaster from the conrete (in less severe cases, partial separation).

6.03 Diagnostic checklist
- *Was the plaster used based on gypsum?*
  BS 5492: 1977: *Code of Practice for internal plastering.* Restricts the choice of plastering to those based on gypsum. Very good adhesion has been obtained by BRE with retarded semi-hydrated plasters.
- *Was a mechanical key provided to the surface?*
  Suitable keys can be provided, in consultation with the plastering contractor, by using shuttering with a rough surface, suitable retarders or special inserts in the shuttering.
- *Was only one coat of plaster used?*
  One coat of plaster is preferable but if two coats are necessary the total thickness should not be more than 9.5 mm. The thinner the plaster, the lower the stresses for a given temperature change.

  To enable one coast or a combined 'thin' two-coat plaster to be applied, the surface of the concrete needs to be level and true and 'fins' are to be avoided. To achieve this, the shuttering needs to be level and true; joints must be tight.
- *Was the surface:*
  1 brushed wet just before plastering but without water being visible on the surface or
  2 treated with a proprietary bonding liquid of the polyvinyl type according to makers' instructions?

These precautions are to help control background suction.

6.04 Further reading
- Building Research Advisory Service Technical Information Leaflet 24, *Plastering on dense concrete.*
  It has been shown that moderate wetting of the plaster improves the adhesion although a good key is the first priority. On very smooth backgrounds PVAC emulsion as a bonding treatment is recommended.
- BS 5492: 1977: *Code of Practice for internal plastering.*
- BRE Digest 213, *Choosing specifications for plastering.*
  Uses a logical sequence of questions and comprehensive tables to facilitate making the choice of plaster.

7 External render

7.01 Causes of failure
Combination of:

- Differential moisture movement of render and background, and
- weak bond between render and background (sulphate attack can also be responsible for loss of adhesion and cracking, the latter especially)

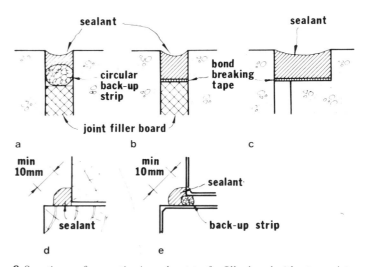

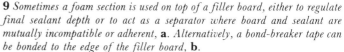

8 *(above) Poor adhesion between a polysulphide sealant and concrete because the latter had been primed with bitumen, thus preventing a chemical reaction at the surface of the mastic shown in the aerated surface,* **a***. Traces of bitumen,* **b***.*

9 *Sometimes a foam section is used on top of a filler board, either to regulate final sealant depth or to act as a separator where board and sealant are mutually incompatible or adherent,* **a***. Alternatively, a bond-breaker tape can be bonded to the edge of the filler board,* **b***.*
Sealants in movement joints must be free to stretch when the joint opens, and not be restricted by adhesion to the back of the joint. Bond-breaker tape, **c***, achieves this. See also pages 129–129.*

7.02 Effects of failure
- Detachment, but more particularly cracking, leading to moisture penetration that may become entrapped and in turn lead to further detachment, cracking or sulphate attack.

7.03 Diagnostic checklist
- *Was the mix for the rendering related to the background material and condition of exposure (see* **7***)?*

Table 1 Properties of sealants

Sealant	Chemical type	Nature	Life expectancy (years)	Movement accommodation %	Max. joint width (mm)	Comment
Hot poured	Bitumen	Plastic	3–10	5–10	50	Available to BS 2499 in 4 types
	Bitumen/rubber	Plasto-elastic	3–10	10–15	50	
	Pitch/polymer	Plasto-elastic	3–10	10–15	50	
Cold poured, 2 part chemically curing	Polysulphide	Elasto-plastic	5–20	15–25	50	Can be pitch modified
	Polyurethane	Elastic	5–20	15–25	50	
	Epoxy	Elasto-plastic	5–20	5–15	50	
Gun applied, non-curing	Oil based	Plastic	5–10	5–10	20	Forms surface skin
	Butyl	Plastic	5–15	5	20	Not recommended for exposed joints
	Acrylic	Plasto-elastic	10–20 +	10–20	20	Available as solvent containing or aqueous emulsion
Gun applied, 1 part chemically curing	Polysulphide	Elasto-plastic	10–20	10–20	20	Primers required on porous substrates. Cure is initiated by atmospheric moisture. Cure rate slower than for 2 part materials
	Polyurethane	Elastic	10–20	15–30	20	
	Silicone	Elastic	10–20	15–30	20	
Gun applied, 2 part chemically curing	Polysulphide	Elasto-plastic	20 +	20–30	25	BS 4254 applies. Primers required on porous substrates
	Polyurethane	Elastic	10–20	20–30	25	Primers required on porous substrates
	Polyepoxide/polyurethane	Elasto-plastic	20 +	50	50	
Hot applied, non-sag	Bitumen	Plastic	5–10	5–10	25	
	Bitumen/rubber	Elasto-plastic	5–10	10–15	25	
Strip	Butyl		10–15	N/A	N/A	Suitable for situations where compression is provided in assembly
	Bitumen/rubber	Elasto-plastic	10–15	N/A	N/A	
	Polyisobutylene/butyl		15–20	N/A	N/A	
Hand applied, bedding compounds	Oil based	Plastic	5–10	Negligible	N/A	Will tend to dry on exposure to air. The life may be extended by painting. Generally unsuitable with aluminium windows

Notes
* Plastic materials exhibit plastic flow and have little or no recovery after deformation.
* Elastic materials (elastomers) have physical properties similar to rubber and return to the original shape after deformation.
* Elasto-plastic and plasto-elastic exhibit partial elastic and partial plastic properties.
1. Movement accommodation is expressed as the total reciprocating movement occurring at the joint. Manufacturers may quote either this or a figure based on the median joint width.
2. Multi-component sealants require mixing prior to application and must be used within their pot life.
3. Elastic and elasto-plastic sealants generally require primers on porous substrates.
4. This table is not an exclusive list of sealant materials available.

10a

Table 2 Suitability of various sealant types

Sealant application	Expansion joint (not subject to traffic)	Joints between cladding units	Pointing of window and door frames	Curtain wall joints	Glazing	Bedding window and door frames	Joints subject to traffic	Water retaining structures
Hot poured							●	●
Cold poured, 2 part chemical curing							●	●
Hand applied, bedding compound					●	●		
Gun applied, non-curing	●	●	●	●				
Gun applied, 1 part chemical curing	●	●	●	●				
Gun applied, 2 part chemical curing	●	●	●	●	●			
Hot applied, non-sag								
Strip					●	●		

10b

10 (right) Facsimile reproduction from DOE, Construction 18, showing the properties of sealants, **a**, and the suitability of various sealant types, **b**.

• Was the background (including previous coats of render) properly prepared before application of the render?
The background surface must be clean and free of all loose particles and capable of providing a good mechanical key to ensure good adhesion.

The suction of the background must be properly controlled — it should be neither too dry (suction increased by dryness) nor too wet (suction reduced but moisture movement increased).
• Was the work carried out in the open during inclement weather?
If the background is uncovered during wet weather (before or after applying the mix) its moisture movement will be increased. Unless the temperature can be maintained above 5°C during the winter frost damage is most likely.
• Was each coat allowed to shrink and dry out for as long as possible before the subsequent coat was applied?
Drying out should not be accelerated.

7.04 Further reading
• BS 5262: 1976: Code of Practice for external rendered finishes (formerly CP 221). The latest reference with detailed advice on all aspects. Includes reference to BRE Driving-rain index.

• BRE Digest 196, External rendered information in BS 5262.

8 Sealants at joints, wall windows

8.01 Causes of failure
• Inability of sealant to accommodate resulting from an incorrect choice of material.
• Inadequate preparation of all surfaces accepting the sealant.
• Degradation of sealant material on exposure, to sunlight especially.

8.02 Effects of failure
• Detachment resulting in water penetration.

8.03 Diagnostic checklist
• Were all surfaces sound (i.e. not friable), clean and dry at the time of application?
The presence of 'foreign' matter on the surface to receive the sealant material invariably results in a weaker bond.
• Were the surfaces suitably primed or sealed before application?
Some sealants such as polysulphides suffer loss of the reacting medium by surfaces with high suction; others may be attacked by the alkali in cement-based products. It is essential that the primer or sealer is compatible with the sealant material when the latter is applied — incorrect priming or sealing materials may interfere with the curing of the surface of the sealant and thus positively prevent adhesion, **8**.
• Was a back-up material used?
The use of back-up materials not only aids adhesion (by ensuring correct depth of sealant material) but also avoids wastage of material, **9**.

The back-up material must not provide restraint to the sealant. If the back-up does not have the same compression or extensibility as the sealant, adhesion is prevented.
• Was the sealant related to the anticipated range of movements?
Sealant materials vary in their ability to accommodate movement and need therefore to be carefully selected, **10**. In addition to the requirements for sealants in BS 4255, parts 1 and 2 and 5215, reliance must be placed on the advice of the manufacturer.

Joint type and width are among important factors that influence the performance of sealants. There may be a large difference between designed and actual joint widths, **11**.
• Was the sealant protected?
Protection against degradation from sunlight is required for many sealant materials. At low level, if unprotected, some sealants may be vandalised; at high level birds may be responsible for extracting the material, **12**.
• Was the sealant correctly applied?
Skill is required in application and poor application affects durability adversely, **13**, see also **14**, pp. 129–130.

8.04 Further reading
• DOE Construction **18**, June 1976, page 4, 'Sealants, mastics and gaskets'.
Good resumé of current availability, properties and uses. Comprehensive list of references.
• BRE Digests 36 and 37, Jointing with mastics and gaskets.
• Manual of good practice in sealant application, Sealant Manufacturers' Conference and CIRIA.

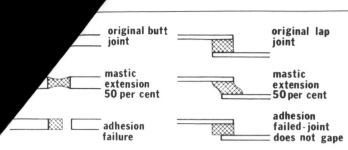

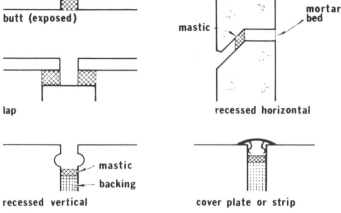

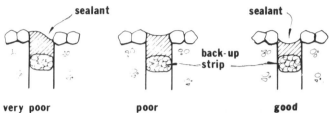

11a *Effect of joint type on mastic seal, and form of failure which may occur.*

12 *Exposure of sealants in joints: protection may be needed.*

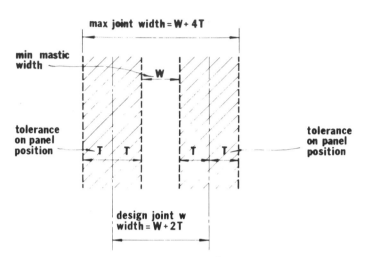

11b *Diagram illustrating design joint width and maximum width.*

13 *The application of the sealant has considerable bearing upon its ultimate durability: diagram* **a** *(from the* Manual of good practice in sealant application) *shows three examples ranging from very poor to good see also pages 128–129*

Somewhat disappointing, considering the source but useful guide to principles. Much repetition, particularly of drawings, which are of poor quality. There remains a need to consult sealant manufacturers about specific applications.

- BRE Information Paper IP25/81, *The selection and performance of sealants* – see update below.
- BRE, 'Selection, performance and replacement of building joint sealants', by J. C. Beech, *Building Technology and Management*, 1983, part 1, **21**(7), pages 23–25 and Part 2, **21**(8), pages 14–16.
- BS 6213: 19??: *British Standard guide to selection of constructional sealants.*
- BS 6093: 1981: *Code of Practice for design of joints and jointing in building construction.* Good coverage of functional requirements and joint types.
- Martin Bruce, *Joints in building.* Wide coverage of principles and practice. Complements Code above.
- *AJ* Series, *Products in Practice* and recently *AJ Focus*. For review of products available. No specific issues noted as coverage changes fairly frequently.

9 External wall tiling and mosaic

9.01 Causes of failure
- Mainly differential movement (chiefly moisture movement) between the finish (tile or mosaic) and background in the absence of movement joints or adequately spaced joints.

   With tiling, use of the wrong adhesive or incorrect use of the correct adhesive.

9.02 Effects of failure
- Detachment

9.03 Diagnostic checklist
- *Were movement joints incorporated?*
   Movement joints 6 mm wide should be provided at storey height and 3 m apart vertically. The joint should be filled with a non-rigid sealant.
- *If a traditional cement-based adhesive was used, was the background properly prepared?*
   The background surface must be clean, free of all loose particles and dry before application of the adhesive. The suction of the background requires control (see 7.03).
- *If a proprietary adhesive was used were the recommendations of the tiling/mosaic and adhesive manufacturer followed?*
   There are no standards or codes for proprietary adhesives.

9.04 Further reading
- BS 5385: Part 2: 1978: *Code of Practice for wall tiling: external wall tiling and mosaics.*
   Tables 2 and 3 on pages 19 and 20 give details of the current adhesives to use on particular backgrounds.

10 Paint to timber windows

10.01 Cause of failure
- Use of wood primers with poor mechanical properties.

10.02 Effects of failure
- Detachment of paint film — blistering and peeling.

10.03 Diagnostic checklist
- *Was the priming coat expected to protect the wood from dampness (i.e. a high moisture content) during the period of several months on site?*

Even if the primer survives the long exposure it is unlikely, with the rather thin coats comonly used, to prevent the timber from reaching a high moisture content. The high moisture content of the timber can lead to blistering and peeling of the paint film (i.e. the whole paint system) later.

Primed joinery stacked under cover but in contact with the ground is certain to retain too high a moisture content.

● *Was the primer applied in a paint shop or factory?*
Primers applied on site are likely to be inferior, as the treatment is normally not thorough and conditions are not usually good. A wide variety of acceptable techniques is available (brushing, spraying, dipping, deluging or flood-coating).

● *Was the primer excessively cheap and quick-drying?*
Common practice but can lead to early failure.

● *Was the primer one of the newer non-lead primers?*
Some of these compare unfavourably with the older (toxic) lead-based materials. Materials conforming to BS 5358: *Specification for low-lead solvent thinned priming paint for woodwork* are at least as durable as the old lead primers and should be used.

10.04 Further reading
● BS 6150: 1982: *The painting of buildings* (formerly CP 231). The painting of joinery is covered specifically in: 25 Wood and 26 Plywood, pages 20–27.
● BRE Digests
261, *Painting woodwork*, 1982 (replaced Digest 106).
304, *Preventing decay in external joinery*, 1985 (replaces Digest 73).

11 Built-up bitumen felt to flat roofs

11.01 Causes of failure
● Incorrect primer or bonding compound or incorrent application of correct primer or bonding coat.
● Vaporisation of moisture which was on surface at time of application; entrapped moisture in roof construction or interstitial condensation. Vaporisation may occur between the layers if moisture is left on the surface of the felt before the bonding compound is applied. Vaporisation of moisture from other sources usually occurs under the first layer of felt.

11.02 Effects of failure
● Blistering or cockling of the felt.

11.03 Diagnostic checklist
● *All layers: was the surface clean and dry before the felt was laid?*
Dust and dirt (most likely on the surface of the deck, of screeds especially) will reduce the bond strength.

Moisture on the surface will not be dried out when the hot bonding compound is applied (as is often mistakenly assumed) but will become entrapped between the layers of felt.

● *First layer: timber structures. Was this secured by nailing?*
● *First layer: concrete or screeded surfaces. Was this bonded overall with a bitumen bonding compound on a properly primed surface?*
Priming of surfaces with a bitumen primer is necessary where a strong bond is required. Apart from assisting in obtaining good adhesion, the primer prevents 'frothing' of the bonding compound.

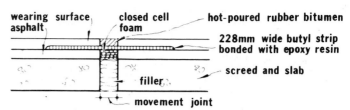

14 *Flush expansion joint demonstrating the use of a butyl strip.*

It is important that the primer is allowed to dry (the coating must be free of volatiles) before the first layer of felt is laid.*

● *Second and third layers: all constructions. Was the correct bitumen bonding compound used? Was the compound at the correct temperature at the time of laying? Was the felt properly rolled?*
The type of bitumen depends on the type of structure and situation; the roofing contractor normally advises accordingly.

To secure an effective bond the temperature of the bonding compound should be in the range of 200°C to 220°C, depending on the air temperature at the time of laying. Overheating of the compound will drive off volatile oils, leaving the residue hard and brittle. In this state it is unsuitable for bonding. If the bonding compound is too cold, partial and completely inadequate bonding results.

The technique known as 'pouring and rolling' is aimed to ensure that the felt is properly rolled onto the hot bonding compound so as to ensure overall bonding. The amount of compound poured or mopped in front of the felt needs to be carefully controlled to ensure that the bonding compound does not become too cold.

11.04 Further reading
● CP 144: Part 3: 1970: *Roof coverings, built-up bitumen felt*, with amendment of March 1978, clause 4.6, 'Technique of laying built-up roofing', pages 22–24.
● BRE Digest 8 (new edition 1970). Includes summary of defects.
● PSA, *Technical guide to flat roofing*, 1987, 3.10, Pages 120ff. Covers all skirting conditions with succinct text and clear (coloured) diagrams.
● Tarmac's *Flat roofing, a guide to good practice*. Clearly written with good diagrams. 3.2 Application techniques, pages 71–74, and 5.1 Built-up roofing detail design, pages 146–159.

12 Asphalt skirtings to flat roofs

12.01 Causes of failure
● Dampness of the background at time of application. Background insufficiently keyed.

12.02 Effects of failure
● Primarily slumping of skirting, but blistering from vaporisation of entrapped moisture is also possible. Slumping of the asphalt may be due to inadequate vertical thickness (i.e. less than 13 mm), softness of the asphalt or lack of surface-reflective treatment.

12.03 Diagnostic checklist
● *Brickwork backgrounds: were all joints raked out or an additional key provided to bricks with smooth surfaces?*

* There is no British Standard for bitumen primer. CP 144 Part 3: 1970 gives details of the characteristics required.

Joints need to be at least 13 mm wide, raked out and brushed clean. The surface of bricks that are very smooth should be hacked or treated with a bonding agent (see 'Concrete backgrounds' below) to provide a key.

● *Concrete backgrounds: was the surface left rough or otherwise treated to form a key?*
Surface treatments include:
1 removal of surface laitance by wire brushing
2 application of a proprietary sand:cement plastics emulsion (normally used as a key for plaster) or
3 application of a proprietary pitch/polymer rubber emulsion.*

12.04 Further reading
● CP 144: Part 4: 1970: *Roof coverings, mastic asphalt*, paragraph 3.6.4, pages 17 and Figs 1 and 2.
● Mastic Asphalt Council and Employers Federation, *Roofing handbook*, May 1980, page 8 and details.
● PSA *Technical guide to flat roofing*, 1987, 3.6, pages 29ff. Very clear guidance with good illustrations.
● Tarmac's *Flat roofing, a guide to good practice*. Clearly written with good diagrams. 5.3 Mastic asphalt detail design, pages 170–177.

13 Impervious sheet floor finishes

13.01 Cause of failure
● Dampness in the background affecting the adhesive.

13.02 Effect of failure
● Detachment of the finish.

---

*Surfaces of lightweight concrete may require the use of expanded metal lathing on sheathing felt nailed at 150 mm centres.

13.03 Diagnostic checklist
● *Was the floor damp-proofed?*
Thermoplastic flooring materials are regarded as finishes that are not necessarily trouble-free but are often laid without protection against (rising) damp. Under severe conditions dimensional and adhesion failure may occur.

13.04 Further reading
● BRE Digest 33, *Sheet and tile flooring made from thermoplastic binders*, April 1963.
● BRE Digest 54, *Damp-proofing solid floors*, new edition 1971.
● CP 102: 1973, *Protection of buildings against water from the ground*.

14 Flush movement joints – asphalt flat roofs

14.01 Basic principles
The traditional upstand movement joint is unsuitable for roofs (e.g. elevated walkways, podia and car decks) that take pedestrian or wheeled traffic.

Proprietary jointing systems using bitumen polymer, neoprene and butyl together with an adhesive compatible with asphalt have been introduced to produce a flush movement joint. One system[†] using a butyl strip is illustrated, **15**. The principles are equally applicable to other systems.

An advantage of these jointing systems is that they can be tailored to suit difficult junctions (e.g. roof and parapet).

For satisfactory adhesion all systems demand strict adherence to the basic principles for the condition of the surface (i.e. dryness and cleanliness) and require the background to have a reasonably flat surface. The edges of the strip used must be well sealed by the adhesive.

† Radmat (London) Ltd, Radmat House, 70 Stonecot Hill, Sutton, Surrey SN3 9HE.

Study 26
Corrosion

1 Location of failures

Failure due to corrosion most frequently occurs in the following locations:

- mild steel reinforcement in concrete (see Section 3)
- fixings (see Section 4)
- water systems (see Section 5)
- cavity wall ties — a regional problem but increasingly important generally (see Section 6)

2 Definition and mechanism

2.01 What is corrosion?
Corrosion is the destructive chemical attack of a metal by agents with which it comes into contact. Destruction occurs as a result of the interaction of a metal with its environment, and this environment includes:

- the atmosphere (for exposed metal surfaces especially), which is usually moist and aggressive in the UK
- dissimilar metals, corroding the metal under consideration where both are in contact with water (could operate either directly or remotely)
- other building materials (in contact either directly or remotely)
- conductive water (i.e. of low pH value and with dissolved matter: particularly relevant to water systems).

2.02 Basic mechanisms of corrosion
Basically theer are two types of corrosion: direct oxidation and electrochemical action. The latter is mainly responsible for the failures included in this study, both ferrous and non-ferrous metals being affected. For convenience a short description of direct oxidation is included; but the emphasis in the following notes is on electrochemical action.

Corrosion by direct oxidation
The formation of a *solid film* on the surface of a metal exposed to air, as the result of corrosion by direct oxidation, usually protects the underlying metal from further corrosion. Solid film formation is more commonly associated with non-ferrous metals and it provides high corrosion resistance, especially to copper. In the absence of solid film formation, as is usually the case with ferrous metals such as cast iron and mild steel, corrosion generally proceeds until one of the reactants has been exhausted.

Corrosion by electrochemical action
Apart from direct oxidation when surface films are formed (see above), corrosion of metals is basically *electrochemical* in nature, that is, a chemical reaction accompanied by the passage of an electric current between an anode (which gains material), during which process the anode is dissolved. The presence of moisture, the electrolyte that enables the electric current to flow, is an essential prerequisite for this type of corrosion. An electric current of this kind, and thus corrosion, may be set up in two quite different situations:

- when *two dissimilar metals* (e.g. aluminium and copper) are in contact with moisture — (see chart **1**); or
- on the surface of a *single* metal (notably a ferrous metal such as mild steel) when, in the presence of moisture, one part of the metal becomes anodic and another part cathodic — usually because either the metal or the electrolyte (moisture) in contact are not uniform.

Note
The products of corrosion occupy a greater volume than the metal from which they are formed. This increase in volume, it is important to note, often leads to cracking of the material in which the corroding metal may be embedded (e.g. mild steel reinforcement in concrete, and fixings in masonry and concrete).

2.03 Further reading
- See notes on BRE Digest 362 above for changes and impending changes.
- Addleson, Lyall and Rice, Colin, *Performance of materials in buildings*, Butterworth-Heinemann, 1992, '3.7 Corrosion', pages 457–541. Updated and revised coverage of all aspects of corrosion including precautions in practice to reduce corrosion previously in Addleson, Lyall, *Materials for building*, Vol. 3, London, Iliffe Books, 1972, 3.07, 'Corrosion', pages 45–87.
- Butler, G. and Ison, H. C. K., *Corrosion and its prevention in waters*, London, Leonard Hill, 1966. A highly specialised book that deals in detail with corrosion in all applications but excluding steel reinforcement in concrete. Very useful reference for corrosion in water suppy, in underground structures and in metalwork structures generally.

3 Mild steel reinforcement

3.01 Effects of corrosion
Common effects are:

- cracking or disruption of the concrete cover
- loss of strength due to reduced cross section of the steel
- rust staining of adjacent areas.

3.02 Causes of corrosion
- Corrosion is usually caused by moisture penetrating the concrete cover to the steel.
- Once the moisture has penetrated the rate of corrosion may be increased because of the presence in the concrete of appreciable amounts of calcium sulphate (gypsum), calcium and other chlorides, or calcium carbonate; or simply because more moisture can penetrate once the corrosion products have cracked the concrete.
- Chlorides that could have been added to accelerate setting of the cement (calcium chloride) or that could have been derived from the sea (notably magnesium

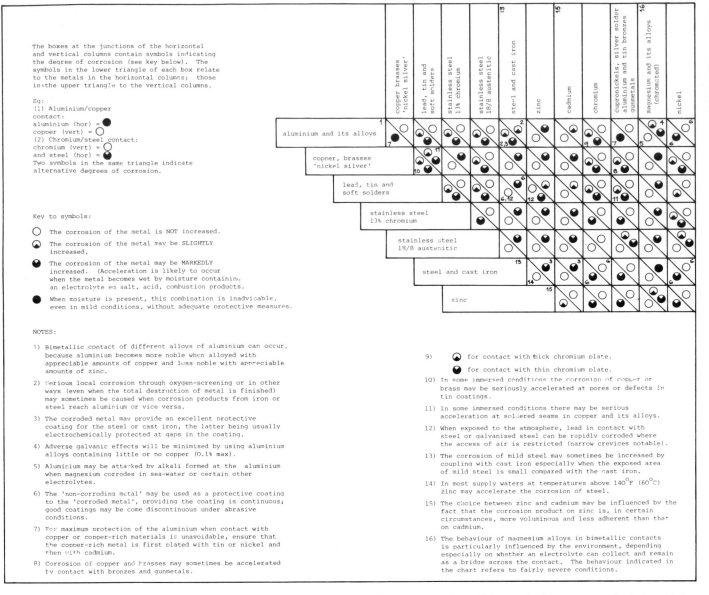

The boxes at the junctions of the horizontal and vertical columns contain symbols indicating the degree of corrosion (see key below). The symbols in the lower triangle of each box relate to the metals in the horizontal columns; those in the upper triangle to the vertical columns.

Eg:
(1) Aluminium/copper contact:
aluminium (hor) = ●
copper (vert) = ○
(2) Chromium/steel contact:
chromium (vert) = ○
and steel (hor) = ◐
Two symbols in the same triangle indicate alternative degrees of corrosion.

Key to symbols:

○ The corrosion of the metal is NOT increased.

◖ The corrosion of the metal may be SLIGHTLY increased.

◕ The corrosion of the metal may be MARKEDLY increased. (Acceleration is likely to occur when the metal becomes wet by moisture containing an electrolyte eg salt, acid, combustion products.

● When moisture is present, this combination is inadvisable, even in mild conditions, without adequate protective measures.

NOTES:

1) Bimetallic contact of different alloys of aluminium can occur, because aluminium becomes more noble when alloyed with appreciable amounts of copper and less noble with appreciable amounts of zinc.

2) Serious local corrosion through oxygen-screening or in other ways (even when the total destruction of metal is finished) may sometimes be caused when corrosion products from iron or steel reach aluminium or vice versa.

3) The corroded metal may provide an excellent protective coating for the steel or cast iron, the latter being usually electrochemically protected at gaps in the coating.

4) Adverse galvanic effects will be minimised by using aluminium alloys containing little or no copper (0.1% max).

5) Aluminium may be attacked by alkali formed at the aluminium when magnesium corrodes in sea-water or certain other electrolytes.

6) The 'non-corroding metal' may be used as a protective coating to the 'corroded metal', providing the coating is continuous; good coatings may be come discontinuous under abrasive conditions.

7) For maximum protection of the aluminium when contact with copper or copper-rich materials is unavoidable, ensure that the copper-rich metal is first plated with tin or nickel and then with cadmium.

8) Corrosion of copper and brasses may sometimes be accelerated by contact with bronzes and gunmetals.

9) ◖ for contact with thick chromium plate.
    ● for contact with thin chromium plate.

10) In some immersed conditions the corrosion of copper or brass may be seriously accelerated at pores or defects in tin coatings.

11) In some immersed conditions there may be serious acceleration at soldered seams in copper and its alloys.

12) When exposed to the atmosphere, lead in contact with steel or galvanised steel can be rapidly corroded where the access of air is restricted (narrow crevices notable).

13) The corrosion of mild steel may sometimes be increased by coupling with cast iron especially when the exposed area of mild steel is small compared with the cast iron.

14) In most supply waters at temperatures above 140°F (60°C) zinc may accelerate the corrosion of steel.

15) The choice between zinc and cadmium may be influenced by the fact that the corrosion product on zinc is, in certain circumstances, more voluminous and less adherent than that on cadmium.

16) The behaviour of magnesium alloys in bimetallic contacts is particularly influenced by the environment, depending especially on whether an electrolyte can collect and remain as a bridge across the contact. The behaviour indicated in the chart refers to fairly severe conditions.

1 *Based on diagram in* Performance of materials in buildings *(see 2.03 for reference). Symbols in table indicate the degree of corrosion risk associated with various combinations of dissimilar materials (symbols in lower triangles refer to metals in horizontal rows; symbols in upper triangles to metals in vertical columns). Thus if aluminium and brass are in contact, the aluminium (lower triangle) is at high risk (black symbol); the brass (upper triangle) is at low risk (white symbol).*

chloride) are common causes of corrosion and can be expected to cause damage in the near future.

- Shrinkage cracking of the concrete can be sufficient to allow the ingress of moisture to initiate corrosion, the rate of which would then be governed by the aggressiveness of the moisture (external surfaces exposed to polluted atmospheres including coastal atmospheres are most vulnerable — see 3.03: 'adequacy in terms of exposure'); by exposure to wind-driven rain; and by the presence of the substances mentioned earlier. Shrinkage cracking is more likely with concretes made with lightweight aggregates.

3.03 Diagnostic checklist: concrete cover to steel

Introduction
Concrete cover is measured from the outside of the reinforcement (including links, stirrups, binding wire, etc.) to the concrete face. The *nominal* cover specified should be as given on the engineer's drawings, but in the context of corrosion failures it is the *actual* cover that matters. The actual concrete cover should nowhere be less than the nominal cover minus 5 mm, and can be measured on site with a cover meter.

Was the designed cover adequate in terms of conditions of exposure?
- The amount of cover required increases as exposure increases and is closely related to the grade of concrete used, **2** (see also next heading).
- Nominal cover for concrete made with dense aggregates should *not* be less than 15 mm for the best-quality concrete, but may have to be as high as 60 mm when conditions of exposure are very severe.
- For concretes made with lightweight aggregates the nominal cover should be increased by 5–10 mm above that required by concretes made with dense aggregates.

Was the designed cover adequate in terms of grade of concrete used?
As **3** shows, the weaker the concrete, the greater the cover required. The requirements for concrete made with lightweight aggregates are not included in the table. These concretes have a lower resistance against gas diffusion and favour carbon dioxide penetration; therefore a nominal cover of 45–50 mm is required for reinforced lightweight

Table 19. Nominal cover to reinforcement

Condition of exposure	Nominal cover				
	Concrete grade				
	20	25	30	40	50 and over
	mm	mm	mm	mm	mm
Mild: e.g. completely protected against weather, or aggressive conditions, except for brief period of exposure to normal weather conditions during construction	25	20	15	15	15
Moderate: e.g. sheltered from severe rain and against freezing whilst saturated with water. Buried concrete and concrete continuously under water	—	40	30	25	20
Severe: e.g. exposed to driving rain, alternate wetting and drying and to freezing whilst wet. Subject to heavy condensation or corrosive fumes	—	50	40	30	25
Very severe: e.g. exposed to sea water or moorland water and with abrasion	—	—	—	60	50
Subject to salt used for de-icing	—	—	50*	40*	25

* Only applicable if the concrete has entrained air (see 6.3.6).

Table 19 gives the nominal cover of dense natural aggregate concrete which should be provided to all reinforcement, including links, when using the indicated grade of concrete under particular conditions of exposure but in addition it may be necessary to specify concrete mix details to provide the required durability (see Section 6).

2 Facsimile reproduction of table 19 from CP 110: Part 1: 1972 (see 3.06 for reference).

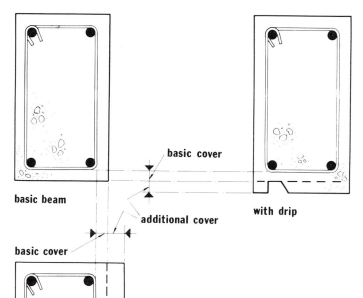

basic beam

basic cover

basic cover

additional cover

with drip

3 *Cover is required for all steel, not just main reinforcement; and additional cover is required where grooves and chases are incorporated.*

4 *Surface cracking to exposed concrete column, betraying corrosion of reinforcement beneath surface. (Photo Building Research Establishment, Crown Copyright.)*

concrete exposed to the weather. Within limits, compressive strength of concrete provides good guidance on the quality and density achieved — generally the higher the compressive strength, the better the quality and the higher the density. Good-quality, high-density concrete provides the best protection.

Was the designed cover adequate in terms of all steelwork?
Cover is sometimes specified for the main reinforcement only, but adequate cover is required to all the steel used including links, stirrups, binding wire, etc., **3, 4***. There should not, however, be excessive additional cover on beam soffits (it could crack).

Was the designed cover adequate for chases, channels, grooves and pockets for drips, fixings and for the thickness of insulation?
These may cause localised reduction in cover.

Was the steel displaced during placing and vibrating of concrete?
Another reason why actual cover to the reinforcement may be considerably less than designed cover.

3.04 Diagnostic checklist: calcium chloride

Introduction
Calcium chloride has been used to accelerate the setting of the cement in concrete for both *in-situ* and precast work. There may be confusion about the advisability or otherwise of its use as there have been changes in recommendations. The use of calcium chloride has not been recommended for some time in prestressed concrete of any kind; more recently its use has been prohibited by PSA in reinforced concrete as well. Earlier recommendations have advised that it should no exceed 2 per cent by weight of the cement; BS 8 110: Part 1: 1985 (see 3.06) now advises that calcium chloride should never be added to any concrete containing embedded metal.

Was calcium chloride used and if so, what was the amount?
See Introduction above. Quantities in excess of 2 per cent by weight of cement should immediately be suspect.

If used, was it added to the mix or to the water?
If added to the mix, the calcium chloride is likely to be unevenly distributed in the set concrete and this could increase the risk of corrosion.

What was the source of the aggregate?
If from coastal areas it could have been unwashed. A chloride would therefore have been added 'unwittingly'.

What about quality of compaction and cover?
Concrete cube tests can give *general* guidance on the quality and density attainable in the area of structure to which they apply, but a *specific* check in areas of difficult compaction may reveal inadequate consolidation, with honeycombing badly repaired. The cover may have been too small for the quality of concrete employed.

3.05 Diagnostic checklist: carbonation of the concrete
The high alkalinity of uncarbonated concrete (i.e. concrete still containing calcium hydroxide) normally inhibits the rusting of the steel reinforcement. Carbonation (i.e. the conversion of calcium hydroxide in the concrete to calcium

* The amount of additional cover shown in **4** could probably be reduced somewhat for sheltered and normal situations, where the full basic cover would not be required behind drips and other recesses.

carbonate by the action of carbon dioxide in the atmosphere) generally proceeds very slowly from the surface to the interior of good-quality concrete; slowly enough to enable steel at the recommended depth of cover to last indefinitely. Carbonation penetrates rapidly in more permeable concretes, i.e. those made with poorly graded aggregates or with a high water/cement ratio. The penetration may be rapid enough to allow carbonation to reach the steel within the lifetime of the building, thus allowing corrosion to proceed. Initiation of corrosion of the steel will be more rapid if the steel has been given less than the minimum recommended cover. Calcium chloride can cause corrosion damage even if the alkali in the concrete has not been neutralised.

3.06 Further reading

British Standard Codes of Practice
- BS 8110: Part 1: 1985: *Structural use of concrete: British Standard Code of Practice for design and construction.* Replaces CP 110: Part 1: 1972.

BRE Digests
- Digest 109, *Zinc-coated reinforcement for concrete*, September 1969. brief outline of mechanism of corrosion pages 1 and 2. Note on calcium chloride page 2.
- 263, *The durability of steel in concrete: Part 1 mechanism of protection and corrosion*, July 1982.
- 264, *The durability of steel in concrete: Part 2 Diagnosis and assessment of corrosion-cracked concrete*, August 1982.
- 265, *The durability of steel in concrete: Part 3 The repair of reinforced concrete*, September 1982.
- 325, *Concrete Part 1: Materials*, October 1987. .Replaces Digests 237 and 244.) Includes comprehensive list of references.
- 326, *Concrete Part 2: Specification, design and quality control*, October 1987. Refers to BS 8110.
- Information Paper IP 6/81, *Carbonation of concrete made with dense natural aggregates.*
- BRE IP 7/89 *The effectiveness of surface coatings in reducing carbonation in reinforced concrete, May 1989.*

General
- Guy I. N., Palmer D. and Parkinson J. D., *Concrete practice*, Cement and Concrete Association, Wexham Spring, 1975, Concise general guidance: page 12, brief note on use of calcium chloride; page 32, table of concrete cover (from CP 110).

4 Corrosion in fixings

4.01 Introduction
Ferrous metal (chiefly mild steel) fixings for cladding seem, on present experience, to be among the most commonly corroded. Fixings for windows may, however, also be included in future. Present evidence suggests that zinc-coated metal fasteners used with some types of copper-chrome-arsenate and ammonium salts based timber preservative treatments may give rise to corrosion problems.

4.02 Effects of corrosion
- cracking of materials in which metals have been embedded
- loss of strength of the fixing or the component fixed
- staining of adjoining surfaces.

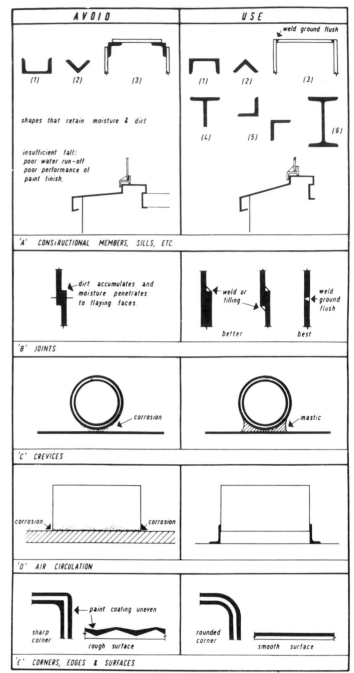

5 *Facsimile reproduction from* Materials for building, *Vol. 3 (see 2.03), p. 83. On left, configurations to be avoided; on right, acceptable alternative.*

4.03 Causes of corrosion
- Insufficient protection of ferrous metals exposed to damp conditions, when *embedded* in other building materials (cement-based products notably) containing corrosive products; or when *fixed in enclosed spaces* subject to water penetration and/or condensation, with inadequate drainage.
- Galvanising, sometimes with the addition of a protective paint based on bitumen, is the most common method used to protect ferrous metals. The sufficiency of the protection, of the galvanising especially, is initially dependent on care taken in cleaning the metal surfaces so that they are free from mill scale, oil, grease and other foreign matter. Thereafter, the thickness and continuity of the zinc coating is important. Components with sharp corners or rough surfaces are particularly vulnerable as the coating may be thinner here than elsewhere. See diagram **5**.

- Equally important, protection can be substantially reduced if the zinc coating is damaged during transit, while being stored on site or during erection.
- Inadequately drained cavities tend to increase the exposure to dampness of the fixings. Similarly, the metalwork itself should be so designed that any moisture collecting on it can be drained off quickly. The risk of corrosion, or the rate of corrosion once it has been initiated, is increased if moisture is allowed to remain on the surface of the metal work or within crevices. This is especially so in enclosed spaces that are also inadequately ventilated.
- Copper-chrome-arsenate salts based timber treatments and zinc-coated steel fasteners under damp conditions.

4.04 Further reading
- GLC, *Development and Materials Bulletin* **91**, 2nd series, January 1976, item 5. 'Corrosion of zinc coated fasteners in conjunction with preservative treated timbers.'

    Report of a test on the resistance of galvanised mild steel punched plate timber fasteners to corrosion by various timber preservatives. Those based on copper-chrome-arsenate and ammonium salts were the most corrosive.

5 Corrosion in water systems

5.01 Type of corrosion
Most commonly, pitting of galvanised components, copper or thin-walled steel tubing in water services (notably hot water systems) and steel radiators in heating systems.

5.02 Causes of corrosion

Type of water in supply and bi-metallic corrosion (galvanised components)
The chemical nature of the water supply should be ascertained or determined by test.

Inadequate cleaning of the inside of copper tubing
If this cleaning is negelcted during manufacture, intense localised ('pinhole') corrosion results in most supply waters. Least trouble has been experienced with tubes complying with BS 2871: Part 1: 1971: *Copper tubes for water, gas and sanitation.*

Air locking within the heating system
This iscaused by entrainment through faulty design or by the generation of gas by corrosion within the system. Air entrainment in the water is also responsible for internal corrosion of thin-walled steel tubing; external corrosion of the tubing results if there is not complete protection by the plastic sleeving (which should be sealed at its ends); and by means of taping of the whole of all joint fittings.

Note
The high temperature of the water in either hot water or heating systems generally increases the risk and rate of corrosion. Where dissimilar metals, **1**, in the system are used in contact with water, electrolytic action and hence corrosion increases with the high temperature. At the very high temperatures used in heating systems (substantially higher than the normal boiling point of water) any salts in solution undergo chemical change resulting in the formation of scale and sludge. Mains water containing substantial amounts of dissolved matter or of a low pH value causes the worst troubles.

5.03 Further reading
- BRE Digest 98, *Durability of metals in natural waters*, revised 1977. Good coverage of all aspects including those faults that commonly occur.
- Bulding Reserch Advisory Service TIL 19, *Pitting corrosion of copper tubing*, February 1974. Draws attention to paragraph 4 of BS 2871 which says that 'the tubes shall be free from deleterious films in the bore' and therefore to the need to use kite-marked (BSI approved) material.
- Building Research Asvisory Service TIL 39, *Thin-walled steel tubing in water services*, March 1971. Emphasises the requirement that this type be used only in closed systems with adequate and joint external protection.
- DOE, *Construction* **18**, June 1976, page 46, 'Water quality in pressurised hot water systems'.

6 Corrosion in cavity wall ties

6.01 General note
At present the corrosion of iron or mild steel wall ties in brickwork appears to be a regional problem confined to those areas where black ash mortar (a mixture of hydrated lime and finely ground ash) has traditionally been used. Similar corrosion can occur with cement/lime/sand mortars if significant amounts of aggressive salts (sodium chloride notably) are present and/or if the ties are retained in damp conditions for long periods (the effects of filled cavities may be relevant in the future). The corrosion process is usually slow but galvanising of the ties does not offer long-term protection.

6.02 Causes of corrosion
The presence of considerable amounts of sulphides in the moist ash increases the rate of corrosion and the presence of chlorides aggravates the corrosion. Together, the low alkaline or slightly acidic conditions produced encourage corrosion of embedded iron or mild steel. Corrosion is generally concentrated on the part of the tie embedded in the outer leaf, **6**.

6a, b *Corroded wall ties. The part of the tie in the outer leaf is usually far more severely corroded than that in cavity or inner leaf. Expansion of corroded part then causes crack in horizontal joint in outer leaf of wall, in which it is embedded.*

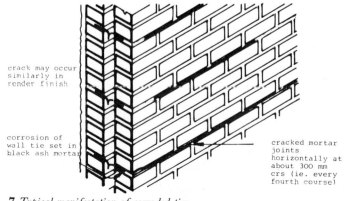

crack may occur similarly in render finish

corrosion of wall tie set in black ash mortar

cracked mortar joints horizontally at about 300 mm crs (ie. every fourth course)

7 *Typical manifestation of corroded ties.*

6.03 Further reading

- Building Research Advisory Service TIL 22, *Corrosion of wall ties in cavity brickwork*, February 1974. Concise explanation of nature of corrosion process, symptoms of wall-tie corrosion and the nature of remedial measures. Better to use later BRE Digests and Information papers listed below.
- BRE Information Paper 28/79, October 1979, *Corrosion of steel wall ties: recognition, assessment and appropriate action.* Deals fully with occurrence, diagnosis and remedial work.
- BRE Information Paper 29/79, October 1979, *Replacement of cavity wall ties using resin-grouted stainless steel rods.* This method is reasonably cheap and easy to use on site.
- BRE Information Paper, IP 28/79, *Corrosion of steel wall ties: recognition, assessment and appropriate action.*
- BRE Information Paper, IP 4/84, *Performance specifications for wall ties.*
- BRE Information paper, IP 6/86, *The spacing of wall ties in cavity walls.* Test results confirmed type of tie required and recommendations in BS 5628 as to spacing.
- BRE Digest 257, *Installation of wall ties in existing construction* deals comprehensively with the problem.

Earlier IP 28/79 is better for diagnosis.
- BRE Digest 305, *Zinc-coated steel* includes coatings on components other than steel. Good for basic principles.
- BRE Digest 329, *Installing wall ties in existing construction,* February 1988
- BRE DAS 19, *External masonry cavity walls: wall ties – selection and specification,* reminds on need for proper type and frequency of installation.
- BRE DAS 20, *External masonry cavity walls: wall ties – installation.* Site application of advice in DAS 19.
- BRE DAS 115, *External cavity walls: wall ties – selection and specification,* June 1988.
- BRE DAS 116, *External cavity walls: wall ties – installation,* June 1988.
- DOE, *Construction* **14**, page 41, June 1975. Strongly advises that zinc coating conforms to BS 1243 whatever the mortar used.
- GLC, *Development and Materials Bulletin* **94**, April 1976, Item 5. Also draws attention to the need for ties to comply with the requirements of BS 1243 (which have been upgraded in 1981). Good photographs of defective coating.

Study 27
Timber decay

1 Location of failures due to timber decay

Decay most commonly occurs in:

- roofs
- walls
- windows and external doors — particularly on the ground floor, and those to humid rooms such as kitchens and bathrooms.

2 Effects of timber decay

The most serious effect is loss of structural integrity.

3 Causes of timber decay

The agency of decay is fungal attack, which in turn is dependent on damp caused by:

- condensation
- rain penetration, at joints opened through movements in particular
- contact with other damp materials, notably brick or block masonry
- contact with very humid air for relatively long periods.

4 Basic mechanisms

Four conditions are necessary for fungal growth, namely:

1 a supply of food material (i.e. wood)
2 suitable temperature (from just above freezing to about blood heat)
3 supply of moisture (moisture content 25–30 per cent) and
4 sufficient oxygen. Poisoning of the food material (i.e. by the use of what are called preservatives) is the only practical way of preventing rot if the moisture supply cannot be controlled (i.e. by maintaining the timber below 20–22 per cent moisture content).

5 Further reading

- Addleson, Lyall and Rice, Colin, *Performance of materials in buildings*, Butterworth-Heinemann, 1992, 'Fungi and insects' pages 422–430 and 'Wood-rots' pages 431–451. Updated and extensively revised coverage of causes and effects of decay and related precautions previously in Addleson, Lyall, *Materials for building*, Vol. 3, London, Iliffe Books, 1972, pages 34–44.
- Cartwright, K. St G. and Findlay, W. P. K., *Decay of timber and its prevention*, 2nd edition, London, HMSO, 1958.
  Specialist coverage but with authority.
- Desch, H. E., revised by J. M. Dinwoodie, Princes Risborough Laboratory, *Timber, its structure and properties*, 6th edition, London, Macmillan, 1981.
  Detailed and specialist coverage but with authority.
- Richardson, B. A., *Remedial treatment of buildings*, 1980
  See appendix 2, page 167 on identification of wood rots, in addition to section 2.2, 'Wood decaying fungi', page 56.
  Thorough description of the different forms of fungus, clearly written, including remedial work required.

BRE Digests:
- 299, Dry rot: its recognition and control, July 1985. Read with 345.
  321, *Timber for joinery*, May 1987.
  345, *Wet rots: recognition and control*, June 1989. Read with 299.
- BRE IP 10/80, *Avoiding joinery decay by design*.
  Summarises recent research and sets out the following guidelines to avoid decay:

1 Use timbers of suitable durability or employ preservative treatment.
2 Any dowels used should be of a durable timber or pretreated with a water-repellent preservative.
3 Design to avoid water traps and horizontal surfaces.
4 Avoid jointed sills and bottom rails.
5 Seal effectively all joints between components.
6 Use durable glues.
7 Seal the edges of plywood.
8 Do not fit mortice locks in the region of a dowelled joint.

6 Diagnostic checklist

6.01 Preservation
- *Was the timber treated with preservatives?*
  Timber containing sapwood must be preserved.*

6.02 Protection from dampness
- *Was the timber (carcassing timber and joinery) adequately protected from exposure to the weather prior to its installation and painting?*
  Timber should be stored indoors or raised clear of the ground and covered with waterproof sheets.
- *Was reliance placed on a primer for protection prior to installation of the timber?*
  Unless of exceptionally good quality (both the primer itself and its application), primers should not be relied upon to provide much protection.
  Moisture trrapped within the timber during painting may result in subsequent blistering of the paint finish.
- *Were the junctions between timber frame and wall adequately designed to exclude moisture?*
  Proper detailing of dpcs and sealants is essential (see Study 22: Rain penetration, 8, page 126 and Study 25: Loss of adhesion 8, page 152).

* The sapwood of all timbers is far less resistant to decay then the heartwood but timbers do differ in the resistance that either their sapwood or their heartwood offer to decay.

- *Was there a risk of interstitial condensation in the roof void or space?*
  See Study 20: Condensation, 4.03.

6.03 Design and fabrication of joinery

- *Were the joints adequately designed and fabricated? Was the sectional size of the timber used large enough and were appropriate adhesives used?*
  Water entry usually takes place at joints that have opened up and modern joints tend to open up easily.
  Flimsy sections for windows and external doors are prone to distort or may be too small to allow proper rebates to be formed. Animal and casein glues are likely to fail in persistently moist conditions.

6.04 Further reading

BRE Digests

- 304, *Prevention and decay in external joinery* December 1985.
  Deals adequately with the aspect considered.

- 261, *Painting woodwork*, May 1982. Updates 106 (withdrawn).
  Covers structure of wood succinctly and factors that influence the selection and durability of the paint finish and preservative treatments. Has useful 'Further reading' list.
- 156, *Specifying timber*, August 1973.
  Useful guide to avoid ambiguity. includes references to moisture content and preservative treatments.
- 286, *Natural finishes for exterior timber*, June 1984.
  useful addition to Digest 261.
- BRE IP 20/87, *External joinery: end grain sealers and moisture control*, December 1987.
- 201, *Wood preservatives: application methods*
- DOE, *Construction* **20**, page 36, June 1978. 'Decay of external joinery'.
  Draws attention to common design faults and alternative details.
- GLC , *Development and Materials Bulletin* **107**, Item 4, July 1977 'Timber decay and infestation — an information note'. A concise summary of the types of decay with a guide to their diagnosis.

Index